Call Center Performance Enhancement Using Simulation and Modeling

Customer Access Management
Jon Anton, Series Editor

Call Center Performance Enhancement Using Simulation and Modeling

Jon Anton
Purdue University

Vivek Bapat
System Modeling, Inc.

Bill Hall
Call Center Services

ICHOR BUSINESS BOOKS
An Imprint of Purdue University Press
West Lafayette, Indiana

03 02 01 00 99 5 4 3 2 1

∞ The paper used in this book meets the minimum requirements of American National Standard for Information Sciences—Permanence of Paper for Printed Library Materials, ANSI Z39.48-1992.

Printed in the United States of America

Library of Congress Cataloging-in-Publication Data

Anton, Jon.
    Call center performance enhancement using simulation and modeling
/ Jon Anton, Vivek Bapat, Bill Hall.
        p.   cm. — (Customer access management)
    Includes bibliographical references and index.
    ISBN 1-55753-182-X (pbk. : alk. paper)
    1. Call centers. I. Bapat, Vivek, 1967–   . II. Hall, Bill,
1944–   . III. Title. IV. Series.
HE8788.A57   1999
658.8´12—dc21   99-41758
                                                        CIP

Contents

Preface

The purpose of this book is to provide readers with an understanding about the role, value, and practical deployment of simulation—an exciting technology for the planning, management, and analysis of call centers. The book provides useful guidelines to call center analysts, managers, and consultants who may be investigating or are considering the use of simulation as a vehicle in their business to responsibly manage change.

By examining current and future trends in the call center and simulation software industries and exploring the synergy that exist between them, the authors present an insight into the elements of successful implementation strategies that involve simulation. The goal is to drive up the efficiency and effectiveness of your business's most visible strategic weapon—your call center.

Acknowledgments

We would like to thank everyone at Systems Modeling Corporation for their dedication and commitment in transforming simulation over the course of several years into a must-have technology that can be deployed for widespread use. Several people from Systems Modeling have contributed to the introduction and success of simulation technology in call centers, including Brad Wyland, Eric Brahney, Glenn Drake, Rienk Bijlsma, Tom Hayson, and Norene Collins.

We are also grateful to Bob Bushey (SBC Communications), Eddie Pruitte (Navy Federal Credit Union), Eric Jacobs (CDS), Katherine Miller (IITRI), Sam Kraut (Oracle Corporation), Charles Maner (Bank of America), and I Ding (UPS) for their enthusiasm and invaluable contributions to this book. As early adopters of simulation technology in call centers, their experiences and leadership will undoubtedly shape the future of the industry. We are also thankful to Marylee Cella for her suggestions and for reviewing draft copies of this book.

In particular, Vivek Bapat would like to thank Dennis Pegden, CEO/President at Systems Modeling, for providing him with the opportunity to participate at several levels in this exciting venture. Dennis' character and leadership are a continual source of inspiration. Vivek would like to especially acknowledge Dave Sturrock, VP of Development at Systems Modeling, for being a source of consistent encouragement. Vivek especially thanks his wife Elizabeth for her support at every step and in all of the things that really matter. Finally, Vivek expresses his gratitude to his parents for their selfless love and support.

Bill would like to thank his wife Linda and two sons, Matt and Justin, for giving him their love and laughter, and, of course, Murray, their Labrador, for his constant companionship while writing this book.

Bill Hall and Vivek Bapat would like to give special thanks to Jon Anton, without whose leadership and support this book would not have been written.

The authors dedicate this book to all the hardworking people in today's call centers. Delighting the customer with outstanding service is a tough and sometimes thankless job. Those who do it so well should be applauded.

CHAPTER

1

The Emergence of Simulation Tools in the Call Center

SWEEPING CHANGES IN TODAY'S BUSINESS WORLD

Driving the sweeping changes in today's business world is the growing awareness that managing customer relationships is a key driver of bottom-line profits. Today's customers put great value on timely access to information. In fact, the vision of the "customer access center" of the future is to make information accessible to customers

- at any time,
- from anywhere,
- in any form,
- for free.

This ease of customer access is fast emerging as the critical element of a global business strategy. In the not too distant future, customers will deal preferentially with those companies deemed most accessible.

Serving as the "lightning rod" for customer interactions, world-class call centers are the single point of contact for customers. According to research conducted

1

at Purdue University, over 50 percent of customer interactions will occur through the call center and the Internet by the year 2000 (Purdue University). Fueled by tremendous advances in the integration of telephone and computer technologies, the call center has the potential for being the company's most potent weapon for maintaining long-term customer relationships.

For many companies, global competition has reduced products to mere commodities that are difficult to differentiate through features, functions, or price. Having reached parity, where price and quality are the "table stakes" of doing business, the paradigm shift is definitely towards customer *accessibility*. Executives are beginning to recognize the potential of the call center as a significant revenue generator, perhaps one of the surest investments they can make in enhancing and creating customer value and bottom-line profits. Return on investments made in customer accessibility is seldom less than 100 percent in the first year and, frequently, even more if customer lifetime value is included in the equation.

In *The 1999 Call Center Benchmark Report for All Industries*, executive managers were asked how they viewed the importance of their customer-service call centers. As can be seen in Figure 1.1, a majority of managers now see the call center as a customer-value management process—not the cost center of the past.

The trend within the call-center industry itself is toward increased complexity. The management and design of the modern call center is becoming so complex because of a number of factors: rapid enhancements in technology (specifically the rampant growth and popularity of the Internet), a myriad of new and demanding customer expectations, as well as reengineering initiatives in telecommunications that include new designs for call routing and staffing strategies devised to satisfy these demands, designs that often exceed them.

Despite this, many organizations still consider their call centers *cost* centers and, as such, are burdened by constant pressure to reduce costs while still maintaining service level objectives. On the other hand, there is a growing trend among

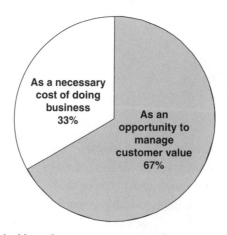

Figure 1.1. How does management view your call center?

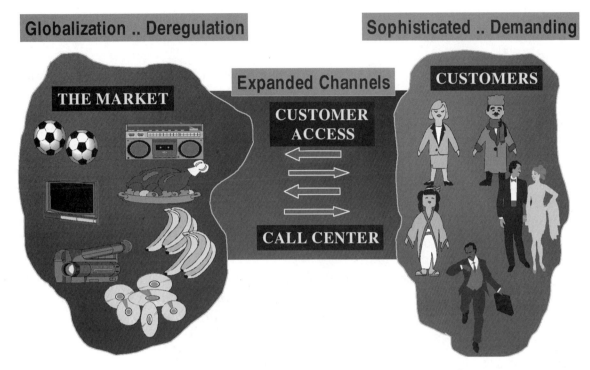

Figure 1.2. Changes impacting the call center

a number of businesses to take the steps necessary to make their call centers into *profit* centers, thereby generating significant revenues for their stakeholders. There are three broad areas of change that affect the modern day call center (Figure 1.2). Let us examine each of these in detail.

1. Changes in the Market
2. Changes in Customers
3. Changes in Access Channels between the Customer and the Market

The effect of external changes (i.e., those involving the market, customers, and the methods of access used to communicate between the market and customers) impacts the internal functioning of the call center in three broad areas.

Work Flow

This area is associated with changes to the work flow of the customer-management process. The changes within these areas may be categorized in two ways: one is outside the call center, the other within the call center.

More than ever before, call centers are now integrated within an organizational structure "outside" their traditional environment. For example, a call center that takes orders over the phone is now tightly integrated into its organization's supply-chain process and has direct connections to vendors, suppliers, manufacturers, and shippers. As a result of this tighter integration, several supply processes can be eliminated as redundant.

Dell Computer is a great example of how a corporation can successfully leapfrog its competition by redesigning and integrating its customer management process with the rest of the corporation. Under the leadership of its chief executive officer, Michael Dell, Dell Computer was among the first computer manufacturers to use the Internet for taking custom orders, and then to link a chain of manufacturers, suppliers, and shippers delivering a customized computer at a price that couldn't be matched by the competition. From corporate and customer points of view, Dell's call center was the trigger that activated a supply chain that was lean from manufacturing to delivery. This transformation of its call center into the link to its outside supply line was the key factor that catapulted Dell to its number one position in the computer industry.

"Inside" the call center itself, processes involving call flow and direction, call handling, interactive voice response (IVR), and scripting (to name a few) are being continuously streamlined and optimized to provide fast and easy access to customers.

Technology

New technology provides call center managers with seemingly endless options for the call handling process. Automatic Call Distributors (ACD) are standard in most new PBXs (private business exchanges). This technology forces managers to determine how the call should be processed. A computer-driven PBX also affords programming capabilities. The Voice Response Unit (VRU) has replaced the agent in many instances. Transactions that once required live agents are now performed through the VRU. These changes make the call-center environment significantly more complex. Vectors can be written and programmed to redirect calls to alternative work groups. The instructions can be time sensitive and are not bound by location.

Virtual call centers, customer-relationship management systems, integration of fax, data, and voice are all key drivers of the technological metamorphosis within modern call centers. The recent benchmark study from Purdue University (cited above) revealed a strong move from business to ride the confluence of the call center and the Internet. According to the 1999 Purdue Benchmark Study, more than 50 percent of customer interaction will occur either through the Internet or call center in 2000.

People

Perhaps the most significant issue on the people side of the call-center environment is determining whether to use skill sets of the specialized agent, or to use universal agents. The Purdue survey showed that more than two-thirds of call centers surveyed still use universal agents as opposed to specialized ones. The survey also noticed a decline in call-center expenditure on agents. The percentage of agent expenditure has dropped from about 70 percent a few years ago to about 56 percent currently.

Agent turnover, a curse of the modern-day call center, is actually dropping. The Purdue Benchmark Report found that turnover was 26 percent for full-time representatives and 33 percent for part-time representatives. These numbers generally rose even higher as the size of the call center increased.

While the report gave average sales per seat per shift as approximately $7,761, the report cited a cost of about $6,300 to bring a new representative on board.

The Purdue Benchmark Report served as a reminder of the volatility and dynamic changes occurring within the call-center environment. What do these types of changes mean to a person involved in call-center management, analysis, and planning? Typically, they boil down to a few key issues, cited in the list below.

1. Balancing staff versus service levels
2. Call center consolidation
3. Implementing simultaneous queuing
4. Determining the best mix of customer-access channels including e-mail, the Internet, and phone calls
5. Implementing virtual call centers through networked ACDs (automatic call distributors)
6. Evaluation of new technology
7. Skill-based routing and scheduling
8. Determining overflow and transfer strategy
9. Designing and validating call routing/handling logic
10. Considering outsourcing services
11. Designing self-service options (IVR [interactive voice response], fax-on-demand, Internet)

The "what-if" capability of simulation technology enables the call-center manager, analyst, and consultant to drive better decisions than those achieved by traditional methods and to virtually eliminate the business risk associated with improper implementation. Simulation permits study of the impact of a proposed change in a controlled environment, without affecting real callers or the "live" call center. Several corporations that have successfully adopted and implemented simulation have saved millions of dollars, have avoided poor implementations, and have driven up their customer management effectiveness to outperform their competition.

Simulation is a proven vehicle for responsibly managing change in your business environment. To understand its value and power to determine practical implementation techniques, and learn first-hand from experiences with simulation in the call-center environment.

OUTLINE OF THIS BOOK

Chapter 1 provides an overview of the major changes taking place in today's call centers. With increasing complexity, new distribution channels, and customers who expect to be "delighted," the challenge for today's call-center manager is to stay one step ahead of this rapidly changing environment.

In Chapter 2, the reader is introduced to simulation (with particular emphasis on "discrete event" simulation). We show how discrete event simulation has the unique ability to model the complexities of call-center systems by including randomness and variability into the analysis process. The chapter also discusses other strengths of simulation, including "what-if" analysis and animation.

Chapter 3 discusses the impact of technology on the call center and how workforce management vendors are adding value to their services by including simulation tools. Other analysis techniques and tools are also discussed, as is the question about why they don't measure up to the power of simulation. We conclude the chapter with a discussion of what can be expected from future simulation products.

Chapter 4 discusses the types of call-center applications for which simulation tools are most effective. Examples are given of tactical and strategic uses and the steps to develop them. The chapter concludes with several examples of simulation modeling approaches.

Chapter 5 discusses trends taking place in the industry and what is driving them from a business and technological perspective. We also discuss the major call-center redesign options available today and how simulation can be used effectively during the evaluation phase.

Chapter 6 describes how to do call center cost/benefit analysis using simulation tools. By forming a statistical relationship between external performance measures (customer satisfaction) and internal measures (such as service level, speed of answer, wait time, etc.) we show how to determine if the call center's current level of service is generating the greatest return on investment (ROI).

Chapter 7 provides important insights into selecting and justifying simulation tools for the call center. We demonstrate how a well-thought-out selection process and implementation plan can have significant impact on the success of your investment in call-center simulation tools.

The book concludes with a series of case studies presented in a question and answer format by actual users of call-center simulation tools. It provides great insight into why simulation tools are being used and how they are being used effectively at many call centers today.

REFERENCES

Purdue University Center for Customer-Driven Quality. 1999. *The 1999 Call Center Benchmark Report for All Industries*. West Lafayette, IN: Purdue University.

CHAPTER

2

The Value of Simulation

EXPLORING THE WORLD OF SIMULATION

To "simulate," according to Webster's dictionary, means to "feign" or to "mimic" a task or activity. There are many different interpretations of the word "simulation." For example, one type of simulation is a computer game such as SimCity. This game allows the user to plan a virtual city or country and to manipulate resources, such as capital, infrastructure, jobs, and utilities. The user can also populate the simulated city with people to reflect the demands and consumption of the city's resources.

In another type of computer simulation, an interactive golf simulator game allows players to choose the type of course they wish to play on and then adapts the visuals and the "feel" of the course according to the players' selections. Virtual or screen images are then used to give players a unique golfing experience tailored to and based on their abilities and skills.

The "Back to the Future" ride at Universal Studios in Orlando, Florida, is an outstanding example of how engineering mechanics can be used to create entertainment through simulated visual stimuli, motion, and sound.

Yet another type of simulation is used in the area of professional training. Flight simulators are used effectively in training pilots, developing their skills to respond to various complex situations, situations which may be too dangerous or expensive to replicate in real life training. For these same reasons, simulated control rooms are also used in training operators in nuclear power plants. At another level, simulated calls are used in call-center management for screening applicants for positions as call-center agents and then in agent training in many call-center environments.

7

In the financial world, various forms of simulation are used—mainly in the form of economic models that capture key elements of market conditions along with a company's financial details for use in predicting market and company performance.

These types of simulations, however, fall outside of the scope of this book. The specific type of simulation technology that we discuss here is called "discrete event simulation." It is important to note that there are many call-center "simulators" available to the call-center industry today. However, instead of using discrete event simulation techniques, many of these tools use mathematical and analytical calculations over compressed time increments in order to "simulate" the behavior of such systems. This type of analysis can be inaccurate and misleading. We will discuss measures to take for understanding and recognizing the differences between such tools and true "discrete event" simulators later on in this book. But first, let's try to understand what we mean by discrete event simulation.

DISCRETE EVENT SIMULATION

Discrete event simulation specifically deals with models and the analysis of systems. Hence, it is especially important to define what is meant by a "system" and a "model" in order to build an understanding of discrete event simulation. Furthermore, it is critical to understand the various relationships between a model and the system that the model represents.

Discrete event simulation focuses on simulating the dynamics of a system in incremental discrete time events. From now on, we will refer to discrete event simulation by the generic term "simulation."

Systems

A "system" can be described as a facility or a process (Kelton, Sadowski, and Sadowski). A system usually consists of some interaction, simple or complex, between the various activities, tasks, and constraints that make up the system. A "system" could be as simple as passengers queuing up at a ticket counter in an airport or a service facility; or as complex as an entire manufacturing assembly line with numerous parts, machines, operators, and material handling devices. An entire airport facility, including baggage handling, flight schedules, agent schedules, and passenger behavior and routing, could also be categorized as a "system." The formulation of a system is completely user-dependent. In fact, there may be several viewpoints of the same system depending on the viewer's vantage point.

Models

As systems become more complex, they pose difficult problems for analysis. The complexity arises because such systems consist of a web of intricate relationships. In order to study a system, it is necessary to implicitly or explicitly define these relationships or to make a number of simplifying assumptions regarding them. In order to measure system performance or to determine how to control systems under varying conditions, many analysts build "models" of the system they want to study.

In some cases, these may be physical models that may be an iconic or a scale model of the real system. These are typically used in architectural settings or in facility design. Given the recent advances in computing, most iconic modeling today is done on the computer, using sophisticated three dimensional (3-D) drawing software.

Logistics or business process problems often require business analysts to use another type of modeling. Such modeling is done through the creation of logic models, which are static in nature and result in a process diagram or a flowchart of all of the activities and processes that make up the system being analyzed. It is important to remember that different users of a systems analysis will require different "views" of the system under study. Process diagramming consists of several methods, including system definition through industry standards such as IDEF (integrated definition language) representations or swim lanes. IDEF is a standard modeling technique created by the U.S. Air Force. Such models are primarily used to better understand the system. Within the broad spectrum of logic models, perhaps the most powerful weapon in the systems analysis arsenal is the *spreadsheet*. A spreadsheet enables users to capture the essential elements of the relationships between various important tasks and interdependencies, which are displayed as equations and formulas. The advantage of using spreadsheets is that they are inexpensive and easy to learn and use, and they provide a good way to conduct quick "what-if" analysis.

RANDOMNESS AND VARIABILITY— KEY PERFORMANCE INFLUENCERS

The problem with most types of systems analysis tools using logical or physical models is that they rarely, if at all, capture essential aspects of the randomness and variability inherent in most systems. Although randomness and variability are arguably the most important parts of a system, surprisingly enough, they are also the most ignored parts of systems analysis.

To illustrate how randomness and variability can have a tremendous impact on the performance of a system (and therefore on systems analysis), let us analyze a

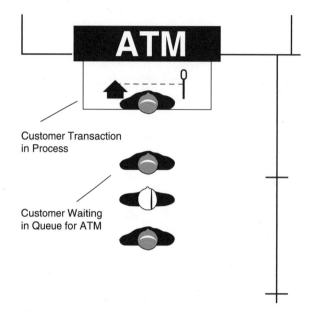

Figure 2.1. ATM processing diagram

very simple system that many of us are familiar with. Let us assume that our system consists of some sort of arrival into a system, a service within that system, and a departure from the system.

Let us assume that customers arrive at an automated teller machine (ATM) for some banking transaction. Let us assume for the sake of simplicity that there is only one ATM available for making transactions. Let us also assume that each customer makes only one transaction. After completing the transaction, the customer leaves the ATM. The next waiting customer can then use the ATM. This is obviously a very simple system (Figure 2.1).

In operations research or queuing theory, such a system would commonly be described as a First-In, First-Out (FIFO) and single server (the server being the single ATM) system. If we were to build a logic model of this system, we would most likely draw a process diagram describing the system as shown in Figure 2.2.

This process diagram aids in understanding and visualizing the system. Let us assume that as analysts, we are trying to make a business decision that hinges on the average expected waiting time for a customer seeking to make a transaction. In order to determine the expected waiting time for arriving customers, let us now add

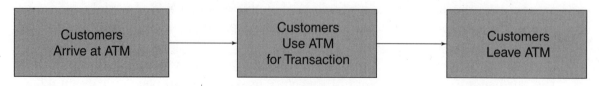

Figure 2.2. ATM process diagram

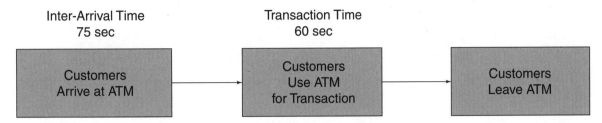

Figure 2.3. ATM process diagram with transaction times

some data to supplement our process diagram. This data may be based on some basic estimates or on historical evidence from similar systems. Let us make the following set of assumptions based on our estimates (Figure 2.3):

- Customers arrive at the ATM approximately every 75 seconds (i.e., an average of 75 seconds).
- Each customer transaction at the ATM takes approximately 60 seconds per customer (i.e., an average of 60 seconds).

If we were to analyze this system using averages, we would conclude that the system would work perfectly, i.e., that the expected waiting time for every arriving customer would be zero. If we use a simple spreadsheet-based calculation, mathematical logic or formulas to determine waiting time (or system bottleneck), we would conclude that there would always be a 15-second "slack" in the system (75 seconds between arrivals and 60 per transaction), such that queues would never build up. Hence, an arriving customer would never encounter any waiting time and would always be served instantly on arrival. We can then summarize our findings as shown in Table 2.1.

This would be true for every customer arriving at the ATM.

The Influence of Randomness

In reality, however, we know that customers don't arrive exactly the same time apart and, in addition, their individual transaction times may be completely different. In reality, there is a random behavior in customer arrival, and variability within the service (or transaction) time associated with each customer.

Table 2.1. ATM findings

Avg. Inter-Arrival Time	Avg. Service Time	Avg. Waiting Time
75 sec	60 sec	0 sec

Table 2.2. Adding randomness to ATM arrival times

Avg. Inter-Arrival Time	Avg. Service Time	Avg. Waiting Time
Random	60 sec	60 sec

Table 2.3. Change in ATM average waiting time

Avg. Inter-Arrival Time	Avg. Service Time	Avg. Waiting Time
Random	60 sec	46.89 sec

Let us ignore the variability but consider only the randomness in customer arrival in our system. If we were to analyze the system based on this new assumption, it is conceivable that two customers might arrive at the same time, and, since there is only one ATM, one of them would have to wait while the other was being serviced. The waiting time for the second customer would be 60 seconds. To summarize, our measure of system performance based on this limited additional data may look like Table 2.2.

If we continued to run our system for a longer period, each arriving customer would experience a different waiting time, depending on the previous arrivals to the system. If we calculated each customer's waiting time and then tabulated the average waiting time for all arriving customers for this period, it would be something other than zero as shown in Table 2.3.

The Influence of Variability

To add more reality to our system, let us stipulate that service or transaction times for each customer will be different, not exactly the 60 seconds originally stated. Let us assume that the first customer's transaction takes more than 60 seconds, say 120 seconds. In that case, our second customer would have to wait for 120 seconds before accessing the ATM. Summarizing our system performance based on these two arrivals, we could then show system performance as presented in Table 2.4.

Again, if we calculated each customer's waiting time and then tabulated the average waiting time for all arriving customers during this period, it would be something other than zero, and would appear as shown in Table 2.5.

Table 2.4. Adding variability to ATM service times

Avg. Inter-Arrival Time	Avg. Service Time	Avg. Waiting Time
Random	Variable	120 sec

Table 2.5. Change in ATM average waiting time

Avg. Inter-Arrival Time	Avg. Service Time	Avg. Waiting Time
Random	Variable	75.23 sec

As we can see, using averages in system analysis can lead to inaccurate results even in the simplest of cases. It is easy to then imagine the havoc that analysis based on averages can have in larger, more complex systems involving variability and a number of interdependencies. This can frequently result in poor decisions and expensive mistakes.

SIMULATION

This is where discrete event simulation comes to the rescue. Simulation offers a unique methodology that incorporate many or all of a system's inherent randomness and variability and, in addition, the architecture to conduct "what-if" analysis through experimentation. One of the key benefits of using simulation is that it is possible to describe randomness and variability in the model just as it occurs in the real system and to include all of the dynamics and relationships between the variables in the system. Let us now take a quick look at how simulation makes this possible.

Referring to our model, let us assume that we have collected some data points on inter-arrival time based on observation of a similar system. If we tabulated our findings, they would appear as shown below:

Inter-Arrival Times (sec):

60
50
45
56
90
45
and so on.

If we represent this information in a graphical format, the occurrences of customer arrival times would look like the graph in Figure 2.4, which demonstrates a spread or distribution of the number of occurrences of specific inter-arrival times in each time bucket.

As we can see from the graph, most of the inter-arrival times fall close to 75 seconds, whereas a few occur closer to the 45-second and 90-second time buckets. Given this information, we can say that the data is spread between 45 and 90 seconds, with most of the observations falling close to about 75 seconds. In order to use this

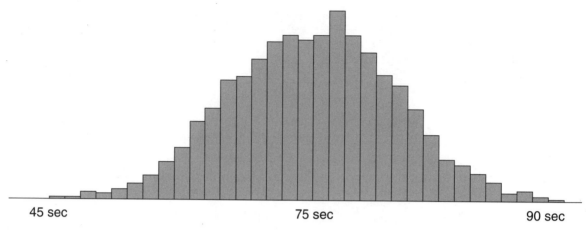

Figure 2.4. Customer arrival times

information in a simulation project, we can use a statistical program to fit a statistical distribution to this data and then determine the best probability distribution that fits our curve.

In our case, the probability distribution that best matches the data was a triangular distribution with a minimum of 45, a maximum of 90, and a mean of 75. In statistical terms, this would be referred to as Triangular (45, 75, 90) as shown in Figure 2.5. For a full discussion of probability, statistics, and distributions, refer to any textbook on statistics.

The simulation program would use this information in creating customer arrivals by randomly "drawing" out samples from this probability distribution for each customer arrival. Most of the samples drawn would be closer to 75 seconds, some might be close to 45 seconds and others closer to 90 seconds. Since the distribution provides limits on the value of the samples drawn, they would never be below 45 seconds nor would they exceed 90 seconds. We must point out that the distributions used for most inter-arrival times are typically exponential distributions that use the *mean* as the key defining parameter.

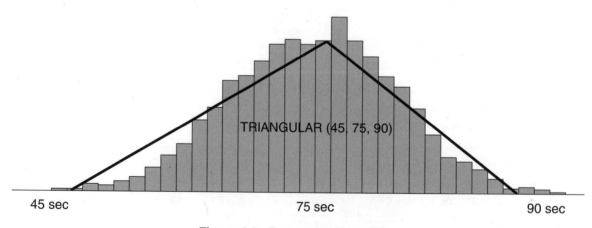

Figure 2.5. Probability distribution

A similar methodology would be used in generating a distribution for transaction times as well. Let us assume that in the case of transaction times, we concluded that we would use a distribution such as Triangular (23, 60, 120). Now that we have addressed randomness and variability in the system, let us use this new information to analyze our simple system in Figure 2.6.

A simulation program permits analysis of the system by "running" a model of such a system over a certain planning period. "Running" the model, in the context of a simulation study, literally means executing the random events that occur in the system over discrete advances in time, just as they would occur in a real system. Each customer would enter the system based on the probability distribution chosen for inter-arrival time and would perform a transaction for a period of time based on the probability distribution chosen for transaction times. The result of running the simulation would give us not only an estimate of waiting times for each customer, but also an estimate of the average waiting time across all customers that have been served. Such a simulation run is typically carried out over a certain period of time—or until some terminating conditions are met. This criterion is determined by the systems analyst.

When we introduce randomness and variability in our systems analysis, we typically need to run several iterations in order to get performance measurements that are accurate within a certain degree of confidence. Statistical theory provides calculations that enable users to determine how many runs to make in order to get statistically valid results. This output is typically measured in terms of degrees of confidence that are user specified. For example, the number of runs to make could be determined by specifying a confidence level (95 percent) in the performance parameters.

Let us assume that in order to get a confidence level of 95 percent or more in our performance estimates, we need to make 20 runs. By running our model 20 times, we have been able to obtain a statistically valid and accurate estimate of expected customer waiting time with a 95 percent degree of confidence. The results are summarized in Table 2.6.

As we can see from the results, what we thought was a perfect system on the basis of using averages in our analysis turns out to be not so perfect after all once we include randomness and variability. The reports show us that the average time spent by a customer is close to a minute (59.813 sec) and that the maximum waiting time is about 4 minutes (247 sec). The ATM was busy during 96.192 percent of

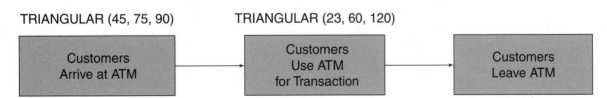

Figure 2.6. ATM process diagram with distributions

Table 2.6. Simulation output for ATM

Run Summary					

Project: ATM Study   **Run execution date: 3/4/1999**
Analyst: Business Analyst   **Model revision date: 3/4/1999**
Replication ended at time: 48000.0

Tally Variables					
Identifier	**Average**	**Half Width**	**Minimum**	**Maximum**	**Observations**
Time Spent in ATM Q	59.813	20.183	.00000	247.00	686

Discrete-Change Variables					
Identifier	**Average**	**Half Width**	**Minimum**	**Maximum**	**Final Value**
ATM Busy	.96192	.01560	.00000	1.0000	1.0000
Number in ATM Queue	.85483	.28209	.00000	4.0000	.00000

the simulation run. In addition, at a certain point in time we had a maximum of four customers waiting in the queue. These results were obtained by studying the arrivals and transactions of 686 customers at and through the system; they are based on a confidence interval of 95 percent. By running a simple simulation model, we have quickly been able to make an accurate and confident assessment of waiting times and to discover a potential bottleneck in our simple system.

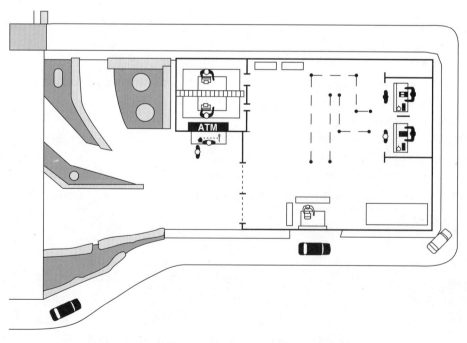

Figure 2.7. Overall banking system diagram

If we had been dissatisfied at the way our system was working, we could now adjust the model to determine what the impact of change to the system would be. Some typical questions that a systems or business analyst would ask would be:

1. What would happen to our service level if we used two ATMs instead of one?
2. What if we installed a faster ATM to replace our current one instead?
3. How would that replacement affect our costs?
4. What would happen if the ATM failed due to some unavoidable circumstance?
5. At what time during the day or month should we schedule routine maintenance so as not to disrupt service?

If we extended our study of the overall banking system that involves ATMs and tellers servicing walk-in and drive-through customers, the system would appear as shown in Figure 2.7, our animated simulation model.

BENEFITS OF SIMULATION

With simulation, we are able to provide answers quickly and accurately to each of the above questions. We are now able to *assess the impact of changes* and make mistakes in the models, preventing costly errors and mistakes prior to implementation. In addition, we can be extremely confident about the quality of our answers since we have included all facets of the system in our analysis. As a primary step, simulation aided our *understanding* of the system, just as any static modeling or flowcharting tool would do. As a secondary step, it also provided invaluable benefits in terms of conducting an accurate "what-if" analysis through experimentation with the model.

By using simulation as a business analysis tool, it is possible to *avoid risk* of error that may only be discovered in the live system at some future state after implementation. Not only is it possible to avoid errors in current implementations, it is also possible to test scenarios that could never be tested in real life because of the associated risks and costs. If an existing system is being enhanced, disruptions in customer service in the working system are avoided by simulating, instead of implementing, changes to an existing system on the computer,

Animation provides yet another powerful dimension to the exploitation of simulation technology. Animation can be used to communicate in compelling ways both the opportunities for change as well as the effect of the proposed changes. Animation also helps to secure "buy-in" from the various management levels associated with a project. Most leading simulation vendors today include animation as a standard component within the software, and many have woven this capability into the modeling process itself.

Today some leading vendors are providing animation in layers. The first layer is a process-level animation which provides a flowchart of the system pro-

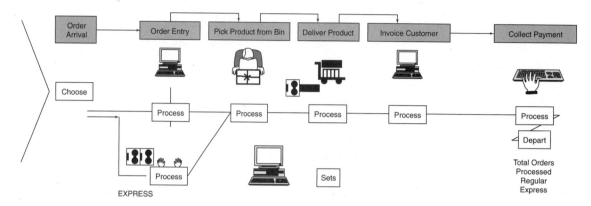

Figure 2.8. Order collection and fulfillment process

cess brought to life by demonstrating the movement of paperwork, customers, or other entities that move through the processes, executing tasks and carrying out activities as shown in Figure 2.8.

Another layer of animation is system-level animation, which allows a look at a facility or a system view, such as a call center as seen in Figure 2.9, a production floor, or a warehouse.

A third-level view provides a combination of both process-level and system-level animation. Each view is focused on a particular user of the model and system.

Another classic benefit of simulation technology is its unique capability to compress or expand time during the course of a run. This enables us to view the state of the system at any point in time, to fast forward through less crucial periods of pro-

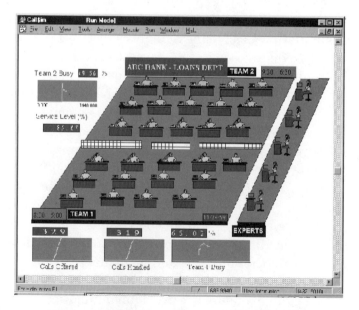

Figure 2.9. System view of a call center

cessing time, while still obtaining a comprehensive view of the entire system. It would be possible to run a simulation of an entire year or more of operation within a few minutes. Or, as in the case of a simulation of the processes of a photocopier, where different tasks and activities take milliseconds, time can be expanded.

We have used a very simple example with the ATM system to describe the enormous impact that randomness and variability have on even simple systems. The system described above is, in fact, very similar to a basic call-center system, where calls come into the call center, queue up before an agent or a self-service device, and then leave the call center.

We are aware that most business problems, particularly those encountered in today's call centers, are not as simple as the system discussed in this chapter; usually a system will include several interdependent processes with random and variable components in each process. The combined effect of the interaction of all these components can be extremely destructive to overall system performance. Many of these systems include significant costs that hinge upon accurate design and analysis. Many of these systems do directly affect the experience of others, including customers, the agents who interact with customers, and call-center managers responsible for coordination of the interactions. If critical components are neglected in systems analysis, expensive mistakes can result. The analysts in today's business climate, responsible for proposing solutions and making decisions, can no longer afford to ignore the central analytic importance of randomness and variability in their analysis. Analysts in today's business climate can no longer afford to exclude simulation as a tool in their analysis toolboxes.

SIMULATIONS TRANSITION THROUGH THE TECHNOLOGY LIFE CYCLE

Let us now turn the clock back a bit to review the transformation of simulation technology viewed as a technology life cycle.

Simulation itself is not a new technology—it was created decades ago. To study the transformation of simulation technology, we can use Geoffrey Moore's characterization of high technology adoption in *Crossing the Chasm*. As we study its transformation, we will discover that simulation has matured over the last several decades into a technology now ready to flow directly into the technology mainstream as well as several niche markets for technology.

Typical of many information technologies, simulation remained the captive of the early adopters, who were mainly working in university and government research programs and agencies. There were several reasons for this phenomenon. In the early days of simulation, a rare and sophisticated combination of analytical and statistical skills as well as in-depth knowledge of a particular systems domain were required in order to harness the power of simulation. The benefits were many, but unattainable by most.

The "innovators" who saw the value of simulation technology were companies undertaking major systems development that involved significant capital expenditures. In the 1950s and 1960s, most users of simulation were scattered around in the aerospace and steel industries and in government organizations. Typically they were part of core groups of highly qualified individuals using software languages such as FORTRAN to write their own simulation languages. This kind of individualized development resulted in proprietary code and single points of ownership, and at the time, simulations had to be run on highly expensive mainframe computers (Kelton, Sadowski, and Sadowski).

During the early 1980s, simulation was entrenched in academia, particularly in U.S. universities and colleges that were interested in expanding their operations research and industrial engineering programs. Courses dealing with quantitative business analysis began to include simulation as part of the curriculum. During this period, simulation's success in the steel and aerospace industries came to be recognized by other industries, but was used only to fix problems that developed in systems after implementation.

During this period, one industry in particular that found success in the use of simulation as a problem-solving tool was the automotive industry. At the time, the automotive industry in the United States was facing stiff competition from Japanese companies, which were extremely efficient and cost-effective in their manufacturing processes. Simulation provided one way for the U.S. automotive industry to determine current bottlenecks in their existing manufacturing lines, to manage resource conflicts, and to streamline operations at the individual plant level. Users of simulation technology were now placed in core groups, typically ranging from one to 15 qualified people, that supplied expert knowledge and analysis to line managers and supervisors through their use of simulation models. Models of complicated production lines were created and implemented by these core groups, decision making was improved, and millions of dollars of cost savings were the demonstrated result of using simulation models. These core groups from industry combined with academicians to form the "technologists" that accelerated the acceptance of simulation technology.

However, it wasn't until the late 1980s that simulation gradually began to make its mark in the business world. During this period, the value of simulation as a predictive tool was widely exploited. Systems began to be designed using simulation techniques prior to, instead of after, implementation. Several large firms as well as several organizations within the U.S. government began to mandate that capital investments above a certain dollar value be approved and justified only after they had been tested via simulation models. These "early adopters" nurtured the growth of simulations technology and rallied support for its entry into mainstream use.

This trend continued until in the early 1990s, when many smaller companies, including consulting firms, began adopting simulation as a powerful value-added tool in their services portfolios and in doing so were able to penetrate many industries that had not until then been exposed to simulation technology. Through such means, simulation began to capture the imaginations of management engineers and business analysts in various fields: logistics, healthcare, the semiconductor and software industries among others. As more and more vendors rushed into the field, successes involving simulation projects were widely advertised and celebrated.

From the early 1990s to the late 1990s, simulation has been transformed into a technology ready to power itself through "the chasm" and break free into mainstream use. Both users and vendors alike are working together to make this happen.

Today, business process reengineering and change management methodologies are quickly emerging as part of a company's standard business regimen (Profozich). Another growing trend is that of customer relationship management. This involves the design of business processes around customer needs in order to foster long-term relationships. Several leading companies that need to maintain a competitive edge are now rediscovering and harnessing computer simulation in order to effectively and responsibly manage such changes in their business environments. It is within these domains (change and customer management) that simulation will find mainstream use.

Early simulation products were generally high priced, difficult to use, and were not well suited to solving mainstream business problems. Over the decades, these products have continuously evolved so that they are now on the verge of breakthrough to the mainstream. Simulation products have now become inexpensive, are extremely easy to use, and are leveraging and providing natural extensions to other mainstream technologies, such as process mapping and flowcharting.

The future of simulation technology lies in achieving two goals. The first is to make decisive inroads into the significantly larger business process-modeling and process-mapping markets; the second is to establish solid successes and growth in the "niche" application domains or vertical markets.

To achieve the first objective, today's simulation tools are now providing more intelligence to static-flow diagrams created using process-mapping tools in order to drive business decisions. For example, a static-process diagram drawn using widely used products such as VISIO or ABC Flowcharter, can now be automatically converted into a live dynamic animated flowchart, which inherits basic process definitions and data with the purpose of enabling decision making and experimentation through simulation. With some simulation packages, these models can easily be scaled up to any level of detail for more system-level use. Rather than restricting simulation to elite (and expensive) simulationists, this trend empowers the business user who is more familiar with the application domain and, hence, can directly derive the benefits of the technology. As shown in Figure 2.10, the process-mapping markets and simulation markets are merging. Simulation vendors are creating software that can interface directly with leading flowcharting software products to

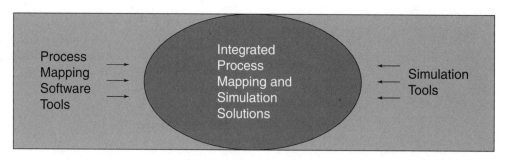

Figure 2.10. Merging of process mapping and simulation markets

leverage the broad-based appeal of process-mapping software, while process-mapping software vendors are adding simulation to their feature list as additional benefits to differentiate them from competing products.

To achieve the second objective, simulation vendors are exploring and exploiting uncharted markets and capturing market share by creating "verticalized" applications focused on a particular industry or a set of problems. By establishing leadership in niche "verticalized" markets, software vendors are able to deliver custom applications to users. Through such specialized applications, users are able to derive solutions to their unique problems without being subjected to the clutter and baggage of unnecessary functionality. This highlights the direction that products are now taking; the emphasis is on reducing learning curves by piggybacking onto broad-based technologies and on the immediate applicability to problem solving by tailoring products to the application domain. Leading simulation vendors have invaded several specialized and growing markets including semiconductor manufacturing, healthcare, and supply chain, where they have delivered resounding successes, making simulation a critical cost saving and analysis tool. The multibillion dollar call center industry represents a significant untapped niche market showing great promise of benefits from use of this technology.

REFERENCES

Kelton WD, Sadowski R, Sadowski D. 1998. *Simulation with Arena*. New York: McGraw-Hill.
Moore GA. 1991. *Crossing the Chasm*. New York: Harper Business.
Profozich D. 1998. *Managing Change with Business Process Simulation*. Upper Saddle River, NJ: Prentice-Hall.

CHAPTER

3

Where Simulation Fits in Call-Center Analysis

THE CASE FOR USING SIMULATION IN CALL-CENTER ANALYSIS—AN EMERGING MARKET

Dynamic and Complex Interaction

A quick look inside a typical call center reveals complex interaction between several "resources" and "entities." Entities take the form of calls—or, rather, customers calling into the call center to receive service. These calls, usually classified by call types, then navigate through the call center according to call-control tables or scripts designed to handle specific nuances associated with each call type. While traversing through the call center, calls occupy trunk lines, wait in one or several queues, abandon queues, and are redirected through interactive voice response (IVR) systems until they reach their destination—the agent or some predetermined self-service destination.

Since the agents have different call-handling skills, it is the customer's request that will determine whether the agent handles the call or transfers it to another agent. Once the call is handled, it then leaves the call center. During all of these transactions, one critical resource is consumed—time. The objective of the call-center manager is twofold. The first is to achieve a high service level, i.e., to get the caller to an agent in the shortest amount of time (measured by waiting time, or in call-center terminology, service level). The second is to provide the caller with the appropriate information in the most efficient manner (measured by call talk time and handle time). The net objective is to minimize the time spent by the caller in the call center while providing the best possible service. These primary measures and objectives usually reflect the performance of a call center.

Balancing these objectives can be a challenging task for call-center analysts. Furthermore, there exists a great deal of sensitivity in the cause and effect of the performance parameters involved. For example, a small adjustment in call routing may have a significant debilitating change on customer service and on the bottom line. A minor reduction in trunk-line capacity may cause too many "busies" and raise the potential for lost customers. Incorrect staffing may cause long wait times, frustrated customers, and exasperated agents. These circular relationships must be defined and analyzed carefully in order to achieve peak performance for the call center.

The Impact of Technology

Over the last decade, advances in technology have brought about many changes in the call-center industry. Undoubtedly, the greatest change has been the private business exchange (PBX). The exchanges were once electromechanical, step-by-step monsters; they have evolved into computerized digital machines with virtually limitless capabilities. Calls are prioritized, transferred, and redirected without human intervention. New technology provides call-center managers with seemingly unending options for call handling. Which is best for the call center: vector-based routing or skills-based routing? Should calls be overflowed to other work groups or should the caller be given the option of leaving a message to be called back?

The call center today that has many different callers with different problems needs to strive to create universal agents—agents that, with the help of technology, can handle almost any caller's request or problem. Figure 3.1 illustrates the popularity of universal agents.

Automatic Call Distributors (ACD) are standard in most new PBXs. This technology forces managers to decide how calls should be processed. Should a call be routed to the agent who has been available the longest? Should agents be allowed to have after-call work? The computer-driven PBX also affords programming capabilities. Vectors can be written and programmed to redirect calls to alternative work groups. The instructions can be time sensitive and redirect calls are not bounded by location.

The Voice Response Unit (VRU) has replaced the agent in many instances. Transactions that once required live agents are now performed through the VRU. How will these changes affect call centers? These technologies are powerful and dynamic, and the effect on business can be tremendous. With all of these changes, evaluation of call centers is more complex than ever before. No longer can man-

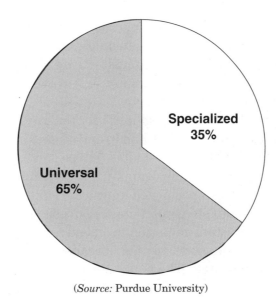

(*Source:* Purdue University)

Figure 3.1. What percentage of agents is trained to be specialized vs. universal?

agers, programmers, or administrators make decisions by the seat of their pants—business is far too critical. But what tool is available that allows decision-makers an opportunity to experiment with technology without fear of impacting their business negatively? The answer is simulation.

HOW THE INDUSTRY IS GEARING UP TO MEET THE SPECIFIC DEMANDS OF CALL-CENTER ANALYSIS

There is now an increasing industry trend among workforce management (WFM) vendors to include simulation tools as a powerful analysis weapon in their arsenal. Simulation provides a key point of differentiation from the competition and offers added value to the vendors' services. This coincides with a growing trend among call center managers demanding that WFM vendors include simulation in their analysis methodologies while configuring WFM systems to suit the unique needs of their particular call center. The combination of these trends is a measure of the growing level of acceptance of simulation in call-center technology.

The simulation industry is gearing up to meet this demand. It recognizes that a WFM system can be a great source for the kind of data needed for a model to provide good results. Not only are the data readily available, but they are now stored in repositories that can be accessed by simulation tools. Advances in simulation technology have made it possible to transfer credible historical and forecasted data, such as call volumes and patterns, agent schedules, and so forth, from these repositories into a simulation model with little or no alteration. Efforts are under way to provide a seamless interface to WFM systems not only from the data entry viewpoint, but also from the reporting side.

In addition, products built specifically for the call-center industry now make it extremely easy to construct a model and to derive useful inferences from them. Modular products with specialized constructs reduce some of the unnecessary features of general-purpose tools, thereby dramatically reducing the learning curve. Something that in the past took specialized knowledge, extensive training, and, usually, weeks or months to complete using general-purpose tools can now be done in minutes or hours. The ability of simulation to embed a call-center submodel deep within a company's supply chain model to study broader organizational issues of strategic importance is an added asset. The path being taken by simulation technologists toward a domain-specific, customizable, and scalable product makes it easier for analysts in the call-center industry to embrace simulation and to rapidly derive benefit from it.

There are several applications for simulation within the call-center industry that provide clear and compelling value over other analysis techniques. Some issues of critical importance to modern-day call centers of all sizes and types are the following:

1. Efficient call handling processes
2. Service level
3. Call center consolidation
4. Skill-based routing
5. Simultaneous queuing
6. Customer abandonment patterns
7. Call routing and overflow
8. Messaging and call return
9. Priority queuing
10. Call transfer and agent conferencing
11. Agent preferences and proficiency
12. Agent schedules

Many of the issues named in the list above could never have been analyzed using traditional methods. In the past, many decisions about procedural change were made based on a combination of best guess, past experience, and rudimentary calculations. Customers were then subjected to disruptions of service while these decisions were enforced. Many poor decisions went unnoticed until after the companies had already paid the high price of lost customers and tarnished reputations. No customer-conscious company can afford to take such risks in this competitive age.

WHY OTHER TOOLS DON'T MEASURE UP

Analytical Tools and Calculators

The most commonly used techniques for call-center analysis are those for staffing and trunking-capacity calculations. A surprisingly large number of these techniques are still based on Erlang calculations (Bodin and Dawson 1996). Erlang formulas were designed in 1917 to solve the question of how many agents would be needed to handle the same number of calls within a single group. The assumptions, listed below, made by Erlang-based analysis make for extremely limited analysis when viewed in the context of today's call centers (Bapat, Mehrotra, and Profozich).

1. Every incoming call is of the same type.
2. Once a call enters a queue, it never abandons.
3. Agents handle calls based on a first-in, first-out (FIFO) basis.
4. Each agent handles every call in exactly the same way.

These assumptions are rarely valid in today's call-center environment. If the individual caller's tolerance for being placed on hold for an agent is exceeded, that caller may abandon the call even if already positioned in a queue; in addition, agents differ in their skill levels and in the time they need to handle various types of calls. The reality is that, in today's call centers, the call requests received are varied in nature, and they may require prioritization and sophisticated call handling to provide better customer service. Yet many companies still base complex staffing decisions on Erlang calculations.

A well-known criticism of Erlang calculations is that they have consistently over-estimated staffing needs. Studies have also shown that 60 to 70 percent of the costs in call centers today are associated with staffing and human resources. This fact, combined with the inadequacies of Erlang-based calculations, can be enormously costly to a call center. Furthermore, it is clear that the application of poor analytic techniques could cause staggering losses—especially when applied to a call center that is growing in size or complexity.

Many spreadsheet-based calculators have improvised on certain aspects of Erlang calculations to patch in more realism in the usage of such calculations for staffing. Some of these spreadsheet-based calculators provide an element of randomness, while others account for some forms of abandonment. However, these patches still do not provide the robustness of the solution provided by simulation. In particular, many industry experts believe that staffing issues associated with advances in skill-based routing and network ACDs can only be studied effectively and accurately through simulation techniques.

Erlang-based calculations are also restrictive and sometimes incapable of analyzing business questions faced by call-center analysts and managers. For example, most of the reengineering done today within call centers involves in-depth understanding and analysis of call flow and process management and not simply identification of who and how many agents should staff the phones. Reengineering issues now pose a whole new set of problems in call-center analysis. In our experience, some of the typical problems that baffle management in an organization that may or may not be directly associated with a call center are similar to the items listed below:

1. How do we best manage our call-center consolidation efforts?
2. How and when should we be overflowing calls to other centers?
3. How should we provide self-service to our customers?
4. What is the most effective call scripting process?
5. Is skill-based routing an option for my call center or for our organization?
6. What changes should we make to allow the call center to handle a new surge of customer transactions through the Internet?
7. How do we resolve issues associated with the customer order and fulfillment process resulting from a phone call or from a web hit to our call center?

Quite simply, the analysis of such problems is beyond the scope of Erlang-based calculations.

Simulation, on the other hand, enables call centers to perform staffing analysis within the framework of a model that allows explicit definition of all the interrelationships among callers, agents, skills, technology, call management algorithms, and techniques (Pegden, Shannon, and Sadowski). This framework ensures the best staffing decisions and provides the analyst with a virtual call center that can be adjusted to answer questions about operational issues and even long-term strategic business decisions. The real value of Erlang-based calculations then comes in providing the initial input data set that is required to run a simulation model.

Operational Tools

Call centers are run today through call management systems (CMS) comprised of workforce management (WFM) systems and switches or automatic call distributors (ACD). These systems not only do the routine tasks of call handling and monitoring, but also provide useful benefits in call forecasting and agent scheduling. Workforce management systems are operational workhorses that manage agent schedules, collect data, and provide detailed reports on how the call center is performing. These systems constitute the backbone of many medium and large call centers. In workforce management, the scheduling of agents is the major driver. Unfortunately, many workforce management systems still rely on Erlang calculations to provide staffing recommendations, and are thus subject to the same limitations as Erlang calculators. WFMs are also limited in their predictive ability and can rarely identify bottlenecks. Furthermore, they cannot address business issues in the way that simulation can.

WHAT CAN WE EXPECT FROM FUTURE SIMULATION PRODUCTS AND THE INDUSTRY ITSELF?

Future simulation products will integrate seamlessly with existing analysis tools in the call-center industry. In earlier sections, we have discussed how general-purpose simulation tools will leverage broad-based technologies such as flow charting and process mapping. Since workforce management tools form the backbone of call-center management, simulation tools used within the call-center market space will, of necessity, be particularly tightly integrated with workforce management tools, and will perform bi-directional data transfer with little or no human intervention. Results of simulation runs will be crunched automatically and available for presentation from every conceivable angle that management may demand.

This trend will change the application of simulation tools from its current strategic use to operational use, thereby reaching a much wider audience and in doing so, crossing the "chasm." Simulation-based products will be placed in the hands of the call-center manager, the analyst, and the business consultant. Simulation engines will become the core of leading workforce management systems and will be used extensively in scheduling agents of varying skill levels and abilities.

More and more, the capital investment decisions made in the call-center industry will be implemented only after they have been validated through the use of simulation models. Call-center vendors and their customers will use simulation models to sell, buy, and configure products and services.

In particular, simulation vendors and workforce management and switch vendors will create partnerships to provide "best-of-breed" solutions to end users. Both technologies have developed independently of each other for several decades. The future will see both of them leveraging each other's core competencies and sharing common boundaries, and, in doing so, they will bring business managers and call-center professionals together to tackle common customer issues.

Internally, both simulation and call management systems will also become all-inclusive. This development will mainly include feature-rich functionality that will enable the call-center manager, analyst, and consultant to better respond and take advantage of emerging technologies and management philosophies that will govern customer management in the future. We have seen examples of this happening already. Many workforce management tools have been upgraded in quick response to the marketplace by providing analysis of skill-based scheduling and virtual call center operations (both new management paradigms), thus improving their feature sets. In a similar move, simulation tools used in call-center analysis are being focused on better integration of back- and front-office processes (such as product fulfillment, field dispatch, repair centers, and contracts) in recognition of the fact that boundaries between the call center and the rest of the organization are dissolving.

The future of analysis tools within the call-center industry is extremely exciting. With the changes brought by the Internet to today's business rules combined with the competitiveness of today's business age, the organization that will succeed will be the one that is forward thinking in its customer strategies, nimble in responding to changing market demands, and technologically savvy in its analysis of those strategies and demands.

REFERENCES

Bapat V, Mehrotra V, Profozich D. 1997. Simulation: The best way to design your call center. *Telemarketing and Call Center Solutions* 16(5):28–128.

Bodin M, Dawson K. 1996. *Call Center Handbook*. New York: Flatiron Publishing.

Pegden D, Shannon R, Sadowski R. 1995. *Introduction to SIMAN Using Simulation*. New York: McGraw-Hill.

Purdue University Center for Customer-Driven Quality. 1999. *The 1999 Call Center Benchmark Report for All Industries*. West Lafayette, IN: Purdue University.

CHAPTER

4

Tactical and Strategic Uses of Simulation Tools

INTRODUCTION

With call-center managers facing increasing challenges to improve call-center performance, powerful new decision-making tools are needed—tools that can effectively visualize, analyze, and improve call-center business processes. The best tools available today to perform these functions are simulation tools, and they work particularly well in the call-center environment. Three major reasons are listed below.

1. **Unpredictable Call Arrival Times**—Calls tend to arrive in bunches at call centers making it very difficult to properly staff without the right analytic tools. Simulation, designed to handle uncertainty, does an excellent job of predicting call arrivals.
2. **Call-Center Complexity**—Too many interactions and interdependencies occur to accurately predict outcomes without using simulation tools. Gut feel and spreadsheets don't work well in this environment.
3. **Volatile Environment**—Most call centers today are going through dramatic changes in technology and work-flow processes.

While the basic structure of call centers remains the same, the rapid changes in call-center technology and the World Wide Web are creating many new ways to deliver service. With these changes comes more complexity, complexity which makes it more difficult to evaluate all of the opportunities available.

31

Because of this complexity, the risk and expense of changing the environment increases dramatically. Imagine being in a workplace burdened by outdated technologies, inefficient work practices, and dissatisfied customers. Changes must be made, but there is the uncertainty about which changes to make or even if certain changes will work at all.

If there were a cost- and risk-free way to assess this outmoded environment without wasting time or disrupting resources, would you do it? Or would anyone choose to maintain the status quo and run the risk of falling further behind the competition? Most of us would choose the former alternative. So the question becomes, "How can we effectively evaluate change while minimizing the risk and expense?" One way to accomplish this is through simulation.

Simulation lets us "model" changes in the environment without the risk and expense of actually implementing the changes. We can try out new ideas for improving call-center processes and work flow, reallocating resources, testing new technology, and changing procedures. By using the powerful "what-if" analysis of a computer simulation model, it is possible, within a calculable time frame, to develop new insights that would be difficult or even impossible to obtain in any other way. There has never been a better time to take advantage of what these tools have to offer.

Once one understands the ability of simulation tools to provide these insights, a whole new perspective on call-center operations is opened up. Secondly, by educating the user about simulation tools, we find that they are used more effectively to solve call-center problems.

As with any tool, simulation tools generally work best when solving a specific type of problem. For instance, while simulation tools won't improve the quality of new hires, they can help in understanding the impact that turnover has on a call center. By identifying the situations where simulation is most effective, precious time is not wasted on the wrong project.

COMMON APPLICATIONS FOR CALL-CENTER SIMULATION

Simulation tools can be used effectively for a wide variety of purposes, some of which are listed below.

1. Perform evaluations
2. Make comparisons
3. Predict outcomes
4. Perform sensitivity analysis
5. Optimize solutions
6. Analyze functional relationships

No matter what type of analysis is being done, the analysis generally begins with a series of "what-if" questions. For instance, what would be the impact on call-center efficiency and effectiveness if:

1. More staff is added
2. A VRU is installed
3. Calls are rerouted to another center after 5:00 P.M.
4. Several call centers are combined into one center
5. Demand increases by 20 percent
6. Calls are routed to the group that has the most expertise in handling them

It quickly becomes apparent that there are a number of questions that can be answered through simulation. Our goal here is not to determine all the questions that might be asked, but to provide the reader with enough insight to develop their own questions and the simulation models to answer them.

We can generally group simulation applications into short-term and long-term planning horizons. The short-term planning horizon is usually reserved for operational issues, whereas the longer-term time frame focuses on major changes and reengineering the way you do business. In this chapter, we will look at examples of typical call-center business issues in both categories and discuss how we could use simulation to analyze and solve them.

Tactical (Short-Term) Applications

Simulation tools are best suited for ad hoc problem analysis. By this, we mean that the tools are not typically used in the day-to-day management of the call-center operation. Instead, simulation tools are generally used when something in the call center is changing or when management is contemplating a change. Let's look at some typical uses:

1. Setting service levels
2. Forecast analysis
3. Allocating staff
4. Budget analysis
5. Customer effectivity analysis
6. Scheduling agents
7. Call-flow analysis
8. Adding new technology
9. Agent proficiency and skill analysis
10. Planning overflow

11. Introducing new products or services
12. Productivity analysis
13. Turnover analysis

Strategic (Long-Term) Applications

Strategic uses of simulation tools typically involve major changes to the call-center operation and may require major capital expenditures. Below are some examples of such strategic applications. Benchmark analysis and consolidation will be explored in more detail later in the chapter.

1. Adding a new call center
2. Consolidation
3. Call-center assessments and benchmarking
4. Becoming a virtual call center
5. Adding electronic access
6. Outsourcing
7. Integration

SIMULATION FRAMEWORK

Before building a first simulation model it is important to have a development process in place. This ensures that the problem is well defined, the right data have been collected, the model is properly designed, and the planned modifications/scenarios are well thought out. There are also "best practices" that should be incorporated into the simulation development process. These can improve the efficiency and effectiveness of the modeling process.

The Modeling Development Process

The basic modeling process includes the eight tasks listed below. By having this or a similar step-by-step process in place, productivity will be much higher and the probability of success much greater. This process can take a couple of hours or less for simple models. If the changes are part of a major reengineering project, the complete process will obviously take much longer.

1. **Identify the Business Issue/Opportunity**—Describe the problem or opportunity facing the call center that you want to solve with the help of simulation. Document why the study is being done and what questions will be addressed. Everyone involved with the project should have a clear understanding of why the model is being built from these statements.

2. **Define the System**—List all the underlying assumptions that are used to define the system. Document the restrictions (constraints) and boundaries that will be used to define the scope and formulate the problem. The goal is to define the system in just enough detail to understand and solve it.

3. **Formulate the Modeling Approach**—Describe the model components, the variables, the processing logic, and the key assumptions. Outline the approach for constructing the model. Determine the measures, metrics, and data to be captured and reported.

4. **Collect and Prepare the Data**—Determine what data is needed as input to the model. Much of this data will come from data resources such as the PBX/ACD, customer databases, other departments, customer surveys, and other sources. Prepare the data in a format that can be used by the model. For instance, some data (such as the average talk time) may have to be run through a statistical program to determine the right statistical distribution fit.

5. **Build the Model**—Using the selected computer simulation tool, build the initial model based on the system description and data defined in the above tasks.

6. **Verify and Validate**—Verify that the model accurately represents the system's current state. Do this by comparing the model to existing reports and reviewing it with people who have intimate knowledge of the call-center operation. This step is critical because all of the "what-if" analysis that follows is based on this model.

7. **Design "What-If" Analysis**—Describe the changes that will be made to the model and the type of output expected to be generated from it. There may be several scenarios (potential solutions) to the problem. All of these should focus on solving the problem as first described above.

8. **Analyze the Results**—Plan on running several cycles of the model for each scenario. Review and compare results to previous runs and project goals until the optimal solution is reached.

Best Practices—Modeling

Collect and Maintain Accurate Data

There is a wealth of information available in PBX/ACD, customer databases, and departments within a company. Decide what metrics are important to meeting the call center's goals and objectives and maintain an electronic version of this information. At a minimum, information will be needed about calls, call routing, and agent staffing in order to get started.

Basic Call Center Modeling Data

Basic call center modeling data will include the three categories of information listed below. Each category gives additional detail about specific types of data that should be gathered.

Incoming Call Information
Call volume for each type of call (over the planning horizon)
Average talk time
Call pattern for each type of call (day of week and time of day)
Call flow (routing logic) for each type of call

Trunk Groups
Trunk capacity
Call types associated with trunk groups

Agent Groups (With the Same Skill Set and Schedule)
Schedule (phone time, lunch, breaks, meetings, etc.)
Which incoming call queues the group serves
Number of agents in each group

Additional Types of Data May Be Necessary

It may be necessary to have certain additional types of data available. The list below gives them by category with specific additional data types.

Agent Groups
Agent adherence to schedule
Agent utilization
Agent turnover rate
Agent costs
Call transfer probability by type of call
Call conference probability by type of call
Adjustments to average length of call by call type

VRU
Average time spent using the VRU
Percent of calls completed at VRU
Percent of call transferred to a live agent

Calls
Call priority (initial plus overrides based on conditions at time of call)
Callback percentages
Calls blocked
Service level
Average wrapup time
Average speed of answer
Average wait time before call is abandoned

Reuse Simulation Models

Simulation models are like spreadsheets. Once developed, they can be used over and over again. Take time to document your baseline models and keep them up to date. When it becomes time to test a new idea, typically 75 percent of the work is done.

Careful Planning and Clear Objectives

Our experience with simulation models shows that without a clear set of objectives, such models can take on a life of their own. They can quickly become too detailed and complicated for solving the problem at hand, or they try to solve too many issues at once. Always remember to keep the model as simple as possible to solve the originally stated problem.

CALL-CENTER SIMULATION EXAMPLES

In this section we have developed several examples of short- and long-term simulation modeling applications. While more in-depth analysis and detail are required for these types of projects, the examples below are intended to give the reader a better understanding of how to approach call-center problems/opportunities using simulation. Remember that, in most cases, there are several alternative methods available to solve a particular call-center problem. The examples used here may or may not be the approach that you would take. Each case depends on the unique characteristics of the call center and potential constraints such as time and cost. In several of the examples below, call center costs, revenues, and customer ratings are contained in the output analysis. These measurements are addressed in more detail later in the book.

Business Issue/Opportunity: Reduce Agent Turnover

Business Scenario

A call center is currently experiencing a 35 percent annual turnover of its agents. The result has been increased talk time and growing customer dissatisfaction. Management's goal is to reduce the turnover rate to 25 percent in the first 6 months and to 15 percent by the end of the first year. Several ideas on how to improve the turnover problem are being considered, but no one knows what the impact will be on the efficiency and effectiveness of the call-center operation. The purpose of the simulation project is to determine the impact on call-center costs, revenues, and performance.

Modeling Approach

Baseline Model

The baseline model provides a view of the operation as it exists today. In order for the model to simulate the turnover problem, two agent groups, "new agents" and "experienced agents," are created for the study. The ratio between the new agents and the experienced agents is based on the current annual turnover ratio of 35 percent. In this model, each group has an equal opportunity of getting an incoming call.

Scenario 1—Lower Turnover Rate to 25 Percent Annually

Once the baseline simulation model was established, two changes were made to reduce turnover to 25 percent.

1. The fully loaded hourly cost per FTE was adjusted to reflect the lower cost of hiring and training.
2. The ratio between newly hired and experienced agent groups was modified to reflect a 25 percent annual turnover rate.

Scenario 2—Lower Turnover Rate to 15 Percent Annually

The same changes made in scenario 1 were then applied to the model in scenario 2, i.e., the fully loaded hourly cost per FTE was further reduced, and the ratio of experienced agents to new agents was increased, reflecting a 15 percent annual turnover rate.

Scenario 3—Experienced Representatives Have Incoming Call Priority

To further improve efficiency and effectiveness the following change was made to the models in scenarios 1 and 2.

If the "experienced agent" group has an agent available at the time a new call enters the queue, *always* route the call to the "experienced agent" group.

Output Analysis

The simulation model generated the following information, as shown in Table 4.1, is based on a one-week planning horizon.

In this example, the number of agents and the number of incoming calls were held constant. By reducing agent turnover in scenarios 1 and 2, the service level, the percentage of calls abandoned, and the average handle time all improved. As these internal metrics improved, so did customer service (as reflected in the "customer rating" index). In scenario 3, customer service and revenue continued to improve as the experienced agent group received more calls. The result of this analysis shows the potential for significant improvement in net revenue by reducing agent turnover.

Table 4.1. High agent turnover

	Baseline	Scenario 1	Scenario 2	Scenario 3
Number of Agents	100	100	100	100
Calls Offered	56,000	56,000	56,000	56,000
Service Level	75	78	84	86
Agent Utilization	65	63	58	56
Percent Abandoned	10	8	7	5
Avg Handle Time	9	8	6	6
Customer Rating	55	61	69	79
Call Ctr. Revenue	$78,000	$85,000	$95,000	$102,000
Call Ctr Cost	$100,000	$94,500	$87,300	$87,300

Further analysis, in the form of a fourth scenario, could explore the impact on the above metrics if the number of available agents was reduced. This reduction may be possible because the call center is operating at a higher level of efficiency (the result of less turnover). This possibility can be explored by adjusting resources until the highest level of net revenue (call-center revenue − call-center cost) is found.

Business Issue/Opportunity— Increased Demand on Call-Center Operation

Business Scenario

A new product and a new advertising campaign to promote an existing product are being launched in the next quarter. These changes will have a dramatic impact on call center demand and the resources supporting it. The purpose of the simulation project is to determine how many additional resources will be necessary to maintain the current level of customer service.

A critical input to this analysis is the forecast of expected demand. The old rule, "garbage in, garbage out" applies here. If the forecast is not accurate, the output from the simulation model will also be inaccurate. A forecasting tool—and not the simulations software—normally generates the incoming call demand. The simulation software does use the forecast to predict the impact on the call-center operation.

To better understand the process, look at Figure 4.1. The forecast feeds the simulation model, which produces output showing the impact on the call center's performance. This information is then analyzed, adjustments are made to the model, and the model is rerun. This process continues until an acceptable level of performance is achieved. Once this process is complete for one forecast, a new forecast can be developed and fed into the model in the same way. In cases where the demand is difficult to forecast, we recommend developing three scenarios—best-case, worst-case, and most-likely scenarios—and feeding each of them into the model. Follow this with an implementation strategy that can support all scenarios.

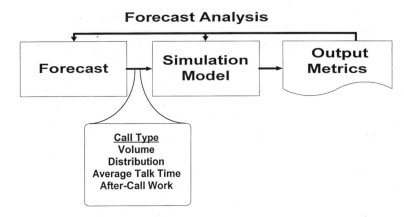

Figure 4.1. Forecast model

Modeling Approach

Baseline Model

The baseline model represents the call-center operation before the new call demand is added. The staffing and performance measures should correspond to results currently achieved by the call center. An illustration of a baseline model is shown in Figure 4.2.

Scenario 1

In Scenario 1 (Figure 4.3), the following changes were made to the baseline simulation model:

1. A new call type (Product C) is added. The call demand for this product over the planning horizon (1 week) is 15,000 calls. The distribution of these calls (by day and within ½-hour increments during the day) is the same as Product A call patterns.

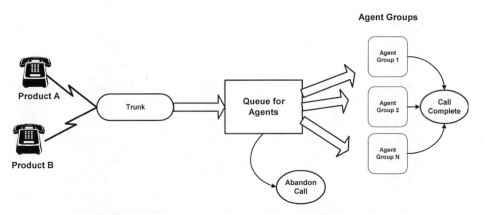

Figure 4.2. Baseline model

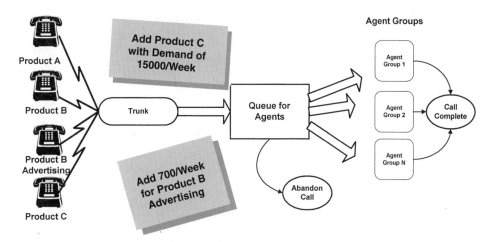

Figure 4.3. Scenario 1 model

2. A new call type (Product B advertising) is added. The call demand for this product is 150 calls between 9 A.M. to 9:30 A.M. and 200 calls between 3 P.M. and 3:30 P.M. This demand occurs only on Tuesday and Thursday.

Scenario 2

In Scenario 2 (Figure 4.4), staffing was increased until key performance metrics came back in line with the baseline model. To accomplish this we added internal staff to support Product C calls and out-sourced Product B advertising calls to another group.

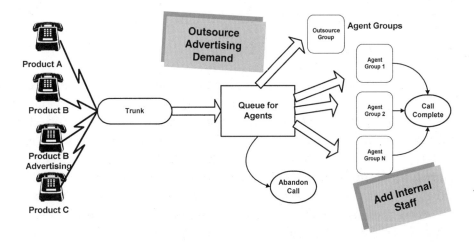

Figure 4.4. Scenario 2 model

Table 4.2. Simulation output—impact of new call demand

	Baseline	Scenario 1	Scenario 2
Number of Agents	100	100	119
External Resources	0	0	30
Calls Offered	56,000	71,700	71,700
Service Level	75	61	75
Agent Utilization	65	87	69
Percent Abandoned	10	16	9
Customer Rating	75	56	79
Call Ctr Revenue	$120,000	$95,000	$149,000
Call Ctr Cost	$100,000	$100,000	$117,000

Output Analysis

The output from the baseline model and the two scenarios (based on a one-week planning horizon) is presented in Table 4.2.

In Scenario 1, the calls offered increased from 56,000 to 71,700. This caused the service level to fall dramatically while agent utilization increased to 87 percent. The number of abandoned calls also increased, and the customer service index fell by 19 points. As expected, adding the new demand without making staffing changes had a significant impact.

To compensate for the new demand, 19 in-house agents were added in Scenario 2, and the advertising calls were rerouted to an outside service. Thirty agents were required to handle these calls. With the additional agents, the service level reached its original level of 75 percent while the customer service index climbed to 79. The results of the study show that while overall costs increased, net revenue improved by $12,000 over the baseline model.

Business Issue/Opportunity—Benchmark Analysis

Business Scenario

A new call-center manager has just assumed responsibility for ABC call center. The call center provides customer support services for insurance policy holders. The manager sees many opportunities for improvement but isn't sure where to start. After some initial research, he decides to do an assessment of the call center by benchmarking it against the "best in class" call centers in the insurance industry. To do this, he obtains a report from Purdue University's Center for Customer-Driven Quality entitled *Call Center Benchmark Report for the Insurance Industry* and begins the analysis.

The purpose of the assessment is to establish a baseline for change. The manager knows that the cost and performance of the call center are critical to its success. The goal is to spend efficiently and perform effectively at a level just slightly better than the competition. The benchmarking process will identify areas where the ABC call center is underperforming in the industry. Once this is established, the manager can develop specific goals and strategies for the call-center processes, the people, and the technologies. Finally, simulation can be used to test different approaches to achieving these goals. The basic modeling process is outlined below. In this example we will focus only on the first three steps.

1. Identify key internal measures that are critical to the call center's success.
2. Develop a baseline simulation model that reflects the current call-center operation.
3. Upgrade the baseline model until key performance metrics match the industry's best in class.
4. Compare the cost structure of this model to the benchmark report's best in class.
5. If the cost structure is below industry standards, develop additional scenarios to improve the cost structure (for instance, consider outsourcing part of the operation, providing for the ability for customers to use the Internet, or rerouting calls to the most qualified agent, etc.).
6. Develop (and build into the model) a revenue formula for the call center based on customer survey data and lifetime customer value (this process is discussed in Chapter 6).
7. With cost and revenue data built into the model, test additional scenarios to find the optimal level of efficiency and effectiveness.

Modeling Approach

Baseline Model

The baseline simulation model was developed from current call-center reports and interviews with key personnel. Figure 4.5 is an overview of the baseline model. Since adherence to schedule was not known, it was initially set at 100 percent. Subsequent iterations adjusted adherence to schedule until the other key measures were in line with current metrics, as shown in Table 4.3.

Scenario 1—Upgrade Performance to Best in Class

Upgrade the existing model until key performance metrics match the goals established in the above table. This will be accomplished by adjusting head count, schedules, and adherence to schedule until the model meets or exceeds the performance requirements. Current processes, technology, and call-center demand will remain the same in this scenario.

Output Analysis

Baseline Model

One of the issues with the current operation was low adherence to schedule. Because the actual rate was unknown, the first model was set to 100 percent for all

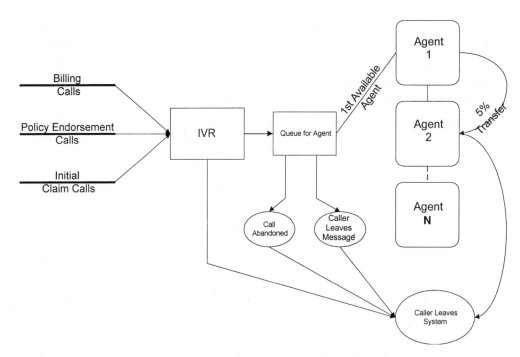

Figure 4.5. Insurance call center model46

agents. This skewed the initial results, showing the call center was performing extremely well, with a service level of nearly 100 percent. It also projected a minimal number of abandon calls with an average speed of answer of 15 seconds and an occupancy rate of less than 50 percent.

Of course, the actual data from the call center tells a different story: A 30 percent abandonment rate, an average answer speed of 45 seconds, an occupancy rate of 60 percent, and absenteeism of nearly three times the industry average. To get the model in line with actual results, the adherence rate was adjusted downward in subsequent runs. When the adherence rate reached 70 percent, the model was accurately predicting call-center performance.

Table 4.3. Performance metrics

Performance Metrics	Current	Goal
Service Level	65% answered in 45 sec	90% answered in 20 sec
Agent Occupancy Rate	60%	85%
Adherence to Schedule	Unknown	95%
Average Speed of Answer	45 sec	15 sec
Average Time in Queue	54 sec	20 sec
Average Calls Abandoned	30%	2%

Table 4.4. Simulation output—benchmark analysis

	Baseline	Scenario 1
Number of Agents	67	45
Agent Occupancy	60	86
Calls Offered/Day	5,000	5,000
Service Level	65	91
Adherence to Schedule	70	95
Percent Abandoned	30	2
Avg Speed of Answer	45	12
Avg Time in Queue	54	16

Scenario 1

The output from Scenario 1 in Table 4.4 shows that the call center needs a minimum of 45 agents to support the current call volume and maintain a service level of 90 percent of calls answered in 20 seconds. While there is much more analysis to be performed, the initial results of this simulation exercise show a very high potential for performance improvement at ABC call center.

Business Issue/Opportunity—Consolidation

Business Scenario

Consolidation provides the opportunity to leverage investments in infrastructure and sharing of resources. By consolidating, a company may be able to take advantage of new technologies that expand call-center capacity and to optimize operational efficiencies.

Consolidation is a major step requiring careful planning and execution. Simulation can play a key role in the analysis phase to quantify benefits before major dollars are committed. Since consolidation can be a very complex undertaking, it may require sophisticated simulation models. However, by taking advantage of simulation tools designed specifically for call centers, many ideas can be tested quickly and easily with minor changes to the model.

To illustrate the modeling concepts of consolidation, let's take a company that has three call centers: one in the Eastern Time Zone, one in the Central Time Zone, and one on the West Coast. The company is planning to upgrade and increase the capacity of the Central Time Zone office and eliminate the Eastern Time Zone office. The upgraded Central Time Zone office will become the primary center. The West Coast center will continue to support its current business and become an overflow and backup facility for the Central Time Zone office.

The purpose of the simulation model is to determine the economic impact of closing the Eastern Time Zone call center, upgrading the Central Time Zone call center, and using the West Coast center primarily as a backup facility. From a modeling perspective, the major activities are outlined below.

1. Create a single simulation model representing the combined activity and performance of the three existing call centers.
2. Reroute the calls from the Eastern Time Zone call center to the Central Time Zone call center.
3. Upgrade performance (people, technology, and business processes) of the Central Time Zone call center.
4. Determine the optimal mix and volume of calls between the Central Time Zone center and West Coast call center.

Modeling Approach

Baseline Model—Single Model of Existing Call Centers

A single simulation model represents the three independent call centers. Within the model each call center is unique. None of the activity of an individual call center has any impact on the other two. Each has its own call types, call volumes, routing scripts, agent groups, performance reporting, etc. By having all the call centers in one model, statistics can be collected across all call centers and summarized into a single report.

The three call centers are illustrated in Figure 4.6. Included in the illustration is a call routing script for each center. Routing scripts are sequences of actions that

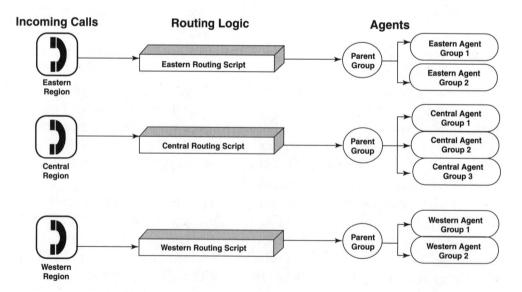

Figure 4.6. Baseline model—call center consolidation

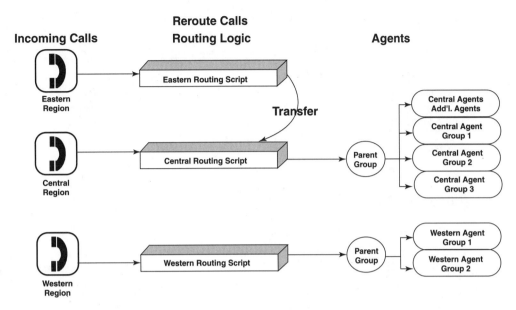

Figure 4.7. Scenario 1—call center consolidation

control the flow of calls through the call center. Routing scripts are very valuable for this type of analysis because they allow calls to be rerouted quickly and easily between call centers and agent groups under a variety of conditions.

Scenario 1—Close the Eastern Time Zone Call Center

This scenario evaluates the economic impact of rerouting calls from the Eastern Time Zone call center to the Central Time Zone call center. There are several ways to do this in the model. One method is to add a "transfer" statement to the Eastern routing script as illustrated in Figure 4.7. This directs all Eastern calls to the Central call center's routing script. As demand increases, Central's performance decreases. To compensate, additional agents are added to the Central group, bringing it back in line with the baseline performance metrics.

Scenario 2—Optimize Calls between the Central and West Coast Centers

After the first consolidation step, scenario 2 looks for ways to improve efficiency and effectiveness between the remaining centers. By making minor changes to the routing scripts, many ideas can be tested very quickly. They can be based on a variety of conditions (that are already built into the modeling software), such as the:

Time of Day

Day of Week

Call Wait Time

Length of Queue

Type of Call

Agent Skill

Agent Availability

Call Priority

Combination of Conditions

Figure 4.8 illustrates this by showing calls being rerouted between the Central and West Coast centers based on changes to the routing script logic.

Output Analysis

Baseline Model

The baseline model represents the starting point for the call-center consolidation process. It provides current performance statistics for each call-center operation. The output in Table 4.5 (based on a one-week planning horizon) shows the West

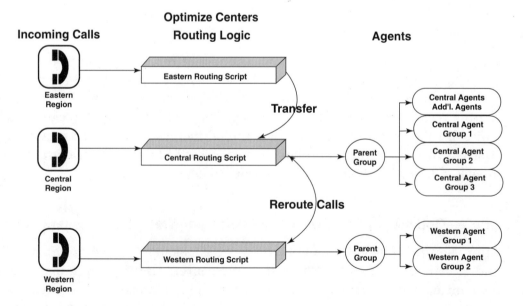

Figure 4.8. Scenario 2—call center consolidation

Table 4.5. Simulation output—consolidation—baseline model

	East	Central	West
FTEs	50	144	83
Calls Offered/Week	35,000	105,000	56,000
Service Level	67	87	92
Agent Utilization	79	78	80
Customer Rating	62	80	85
Call Ctr Rev/week	$45,000	$145,000	$95,000
Call Ctr Cost/week	$50,500	$124,000	$77,000
ROI	−10%	17%	23%

Coast call center operating at the highest level of the three, with the East Coast call center operating at a loss.

Scenario 1

In Scenario 1, as illustrated in Table 4.6, the East Coast center calls were rerouted to the Central call center. To compensate for the increased demand, additional agents were added to the Central call center. The results show a net improvement in net income and a reduction in FTEs.

Additional Scenarios

Because of the many alternatives available for rerouting calls between the two remaining call centers, we did not develop specific outputs. The important thing to remember is that once the baseline model is completed, a whole series of "what-if" questions can be addressed very quickly by making simple changes to the routing scripts as well as other model components.

Table 4.6. Simulation output—consolidation—Scenario 1

	East	Central	West
FTEs		186	83
Calls Offered/Week		140,000	56,000
Service Level		85	92
Agent Utilization		80	80
Customer Rating		78	85
Call Ctr Rev/week		$184,000	$95,000
Call Ctr Cost/week		$160,000	$77,000
ROI		15%	23%

SUMMARY

Simulation can be used to analyze, evaluate, and solve many complex call-center problems. By applying a consistent and disciplined approach, simulation will produce superior results. The five key items to remember are listed below:

1. **Prepare a plan**—Have a clear statement of goals and objectives.
2. **Collect data**—Collect and verify enough data to adequately understand the problem and the current operation.
3. **Develop baseline model**—Develop and verify a baseline model that accurately reflects the real-world operation.
4. **Perform "what-if" scenarios**—Test ideas to improve efficiency and effectiveness.
5. **Verify results**—Make sure the decision-makers have confidence in the model's results.

REFERENCES

Purdue University Center for Customer-Driven Quality. 1999. *The 1999 Call Center Benchmark Report for All Industries*. West Lafayette, IN: Purdue University.

CHAPTER

5

How Simulation Embraces Future Call-Center Trends

BACKGROUND

Over the next few years more and more companies will be moving their low-technology call centers from their back-office support position to the front line of the enterprise as a profit center. Supporting this migration is new computer and telecommunications technology for both voice and data applications. As new state-of-the-art technology is installed, the customers gain more flexibility on how, when, and where they can access the company. While voice is still the dominant means of communication, electronic access by fax, e-mail, the Internet, and kiosk is growing rapidly.

Driving this call center development is the growing awareness that managing customer relationships has become key to creating and managing value in the enterprise. Today's customers want timely accessibility. They want that access at any time, from anywhere, in any form, and they want it for *free*. The customer access call center is becoming the focal point of this interaction. What then does the typical telephone agent know about the caller before answering the telephone? See Figure 5.1 from the Purdue University Benchmark Report.

If managing the customer relationship is key to long-term growth and profits, then designing and implementing a call access center to support these needs is critical. In the past, call centers were created for different reasons. They typically got

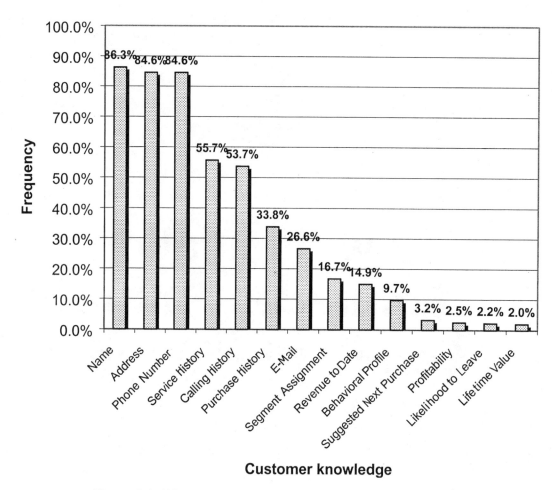

Customer knowledge

Figure 5.1. What customer knowledge do TSRs have access to?

started by supporting new products or services. Professionals who had knowledge and expertise in a specialized area of the business staffed them. Over time, this resulted in specialized call centers being created to support different aspects of the business. As companies began to realize the benefits of sharing customer information with other departments and providing the customer with a single point of contact, the transformation began. While moving the call center from a *cost* center to a *profit* center is a major undertaking, it can have very positive results. If successful, the company and the call center will ultimately become one and the same.

In this chapter, we will discuss the trends taking place in the industry that are driving the changes; we will discuss them both from a business and from a technology perspective. While new technologies make it easier for the customer to do business with the company, at the same time providing much-needed data for the enterprise, the underlying infrastructure supporting these endeavors becomes more

Table 5.1. Call center performance metrics

Metric Description	Inbound Call Center Service Level Statistics		
	Median	**Average**	**Standard Deviation**
Speed of Answer (sec)	25.0	36.0	40.5
Talk Time (minutes)	3.3	4.3	3.2
After-Call Work Time (minutes)	1.2	3.2	8.8
Calls Abandoned	4.3%	5.4%	5.1%
Time in Queue (sec)	30.0	49.7	60.8
Calls Closed on First Contact	83.0%	78.3%	20.3%
Percent Calls Blocked	4.84%	7.97%	10.23%
TSR Occupancy	80.0%	76.0%	16.7%
Time Before Abandoning (sec)	50.0	69.5	75.7
Adherence to Schedule	92.0%	90.2%	9.6%
Percentage Attendance	92.0%	86.2%	19.7%
Inbound Calls per 8-hour Shift per TSR	65.0	74.6	64.8

(*Source:* Purdue University, pp. 5–10.)

and more complex. If this infrastructure is inflexible and cannot be integrated with all aspects of the business, the enterprise will find it increasingly difficult to respond to future challenges. Obviously, making the right decisions is critical for the long-term health and prosperity of the enterprise.

Table 5.1 shows some of the statistics tracked by the annual Purdue Benchmark research.

As companies embark on major structural changes to improve the call center metrics shown above, they can reduce the risk by using simulation tools during the analysis phase. Simulation is able to do the following:

1. Support the evaluation process without disrupting the day-to-day operations.
2. Test the feasibility of certain hypotheses.
3. Provide insight on how certain components will react with each other.
4. Test new situations in order to explore theoretical future events.
5. Answer numerous "what-if" questions that are strategic to the enterprise.

In almost all cases, the simulation process begins with an "as is" model of the current operation that establishes a benchmark for the "what-if" analysis. The "what-if" analysis looks at several business scenarios that might be considered as alternatives to current practices. For each scenario (depending on the type of changes to the model), a comparison of key metrics (like those listed below) is made to the "as is" model. Each scenario is then analyzed to determine the impact of these changes.

1. Speed of answer
2. Time in the call center
3. Talk time
4. Abandonment percentage
5. Service level
6. Costs and revenues
7. Trunk utilization
8. Agent utilization
9. Customer service index

Because we expect call centers to continue the rapid growth, the increases in complexity, and the constant innovation for the foreseeable future, simulation tools will play an important role for those with the responsibility for improving call-center performance.

CALL-CENTER REDESIGN OPTIONS

There are several approaches that businesses can take to improve call-center performance. They range from optimization of what is already in place to a strategy of complete replacement. Each company has to determine the best approach for itself by considering such things as its long-term customer relationship strategy, its competition, and its current structure and resources. Today, companies are doing more outsourcing and consolidating more call centers than ever before. They are also moving toward multifunctional call centers that combine activities such as sales and customer service. Here is a look at some of the options available. The options discussed here are not mutually exclusive. A call center could implement multiple options or make short-term improvements while at the same time designing and implementing a major transformation.

Optimization

Optimization is a route for companies to take whose stated mission is already on track. In such a case, company goals can be accomplished by simply optimizing the existing call-center model. Using this approach means looking for new ways to do it better, faster, and cheaper than the call center is already doing.

But because optimization simply focuses on improving the efficiency of the current operation, this is usually not the right strategy for most call centers to embark on. Today, the majority of call centers need to perform major rework. Nonetheless, even for those call centers already operating on target, simulation tools can provide an excellent means for improving what is already in place.

Consolidation

Consolidation is an option for companies that need to integrate or consolidate multiple call centers into a single homogeneous unit. Threshold size of a center and its degree of specialization are important factors impacting call-center consolidation. Consolidation offers the advantages of leveraging infrastructure investment and the sharing of resources. Call-center consolidation can be managed physically, or it can be managed logically through networking.

Shared Services

The shared service approach is an option for companies with separate and distinct call centers (each with different business functions) that want to share investments in equipment, systems, and people. It brings the agents from all business functions under one roof, allowing them to gain exposure to the company's entire product line and customer interactions. Unlike consolidation, a shared-services approach is conceived from the ground up to serve the entire company's needs.

Distributed Environment

Today most companies have multiple call centers distributed among several locations. Local work force considerations, technology, avoidance of service disruptions due to natural disasters, and costs are a few of the considerations that affect the distribution of call centers. The movement now for most companies is toward consolidation rather than continuation of the trends that led to center distribution.

Outsourcing

Outsourcing one or more functions to a service outside the company may offer several advantages. Several situations that can point a company to the potential advantages of outsourcing are listed below:

1. A company may lack certain skill sets needed to accomplish a strategic objective.

2. A company may be in a highly volatile call-volume market where trained specialists are needed to handle periodic overflow calls.
3. A call center may have high internal costs for relatively simple calls.

Outsourcing can also speed implementation of a function because the key components and infrastructure are already in place at the outsource agency. Today, many companies are considering outsourcing, and for them a key consideration is whether the function supported by the call center is a core competency of their business.

Transformation

Transformation is a strategic undertaking for any company to make. It requires stepping back and repositioning the call center as the single point of contact for the customer. The call center's mission and goals are transformed, moving it from a back-office *cost* center to a front-line-strategic *profit* center. Such transformations represent a major change in the way the company does business.

TECHNOLOGY INFRASTRUCTURE

New computer and telephone technology is having a major impact on call-center operations. Many of the options discussed above are driven by new technology, but the design and implementation of a technology infrastructure is beyond the scope of this book. However, we can show how simulation is used to evaluate technology alternatives.

Let's take an example:

Plans are under way to implement a sophisticated voice network that will consolidate several call centers into a single operation. The new technology allows calls to be routed or rerouted to any call center in the country under a variety of rules and conditions. While "gut feel" and vendor data say that networking the call centers is the right approach, there are no hard facts to prove increased efficiency and customer satisfaction. In this case, simulation is the only analysis method that can accurately model the complexity of a networked call-center operation.

The analysis begins by building a model of the current call-center environment. The model contains multiple call centers, each operating independently. This becomes the baseline model. With the goals and objectives of the study firmly in mind, a second model is developed that incorporates the features and functions of a networked environment. Next, several scenarios are developed to test the features and functions of the new environment. After each scenario is run, the model's performance metrics are analyzed. When all the scenarios are completed, the model that meets all of the objectives and maximizes performance is chosen. This model is compared to the baseline model. If the networked model shows significant improvement in performance, then we have established, with a high degree of confidence, that the new technology makes good business sense.

When building a simulation model to evaluate new technology, a list of the new technology's features and functions is the place to start. Next comes consideration of how each new feature and function can improve the call center's operation. For instance, consideration will be give to whether a particular feature or function reduces talk time, frees up agents, or reduces callbacks, etc. These improvements can be incorporated into the model to evaluate their impact on call-center performance. The evaluation results can then be used to support management in making the key decisions that drive call center expenditures.

Today, the new technologies permit many different call-center functions. Several of the major call-center functions available to companies are listed below. Simulation can be used to evaluate the cost effectiveness of these new technologies before the expenditure is made.

Networking Call Centers

Logically networked call centers (networked to achieve consolidation) require an in-depth understanding of call flows between the company's sites and the cost components of the network. Procedures, agent skills, data, and equipment must also be taken into account to be sure consistency is maintained no matter which site takes the call.

Sharing Customer Information

A common database can be shared by multiple groups within the enterprise. This allows information obtained through the customer service center to be leveraged by other units in the enterprise. Such customer information, based on individual customer profiles, can also be used to route a call to the right agent.

Providing Electronic Access

The movement from voice-only access to electronic access via such technologies as the Internet, electronic data interchange (EDI), e-mail, and other technologies is becoming more prevalent. Technology not only offers new services to the customer for accessing the company, but it can also link the customer to revenue-producing services. But allowing customers to access the center in whatever manner they choose requires major changes in call-center design and operation. Costs, service priorities, service levels, transaction processing, staffing, and other considerations must all be taken into account when planning for electronic access.

Integrating Voice and Call Processing

Integration of voice and call processing (hardware and software components) is necessary to facilitate the concept of "one-and-done" (satisfying all the caller's needs at the time of initial communication). To achieve one-call servicing, a call is linked to other applications and other systems (such as the customer information file, expert systems, and fulfillment applications) in order to meet all the caller's requirements in a single call. Much of this linking of systems and applications will be enabled by computer-telephony integration (CTI).

Routing Calls

Skill-based routing matches the caller's needs with an agent's skills. The purpose is to leverage the agent's skills, time, and value to the customer into a means for improving call throughput and customer satisfaction. Business routing is even more sophisticated. It attempts to determine the caller's needs, value, relationship, etc., during the routing process in order to improve the chances of a successful call. Blended (inbound and outbound) routing is another way to improve agent utilization by switching between inbound and outbound calls as demand fluctuates.

EMERGING TRENDS

Managing customer relationships is the key to long-term growth and profitability. This is driving the development of call centers, which are rapidly becoming the *single point of contact for all customers*. The implications of the call center becoming the single point of contact with their customers are, for most companies, enormous. It means providing access in any form at any time from anywhere the customer chooses. Secondly, it means an integrated network of voice and data transactions that can be accessed anywhere in the company.

Because of the complexity involved in making major changes to call center operations, simulation is the only tool, in our opinion, that can accurately predict performance outcomes. To give the reader a better idea of how this can be done, we have listed six trends occurring in call centers today. For each trend we outlined (see Table 5.2) a high-level simulation approach and a list of questions that could be ad-

Table 5.2. Using simulation to analyze emerging trends

Trend	Simulation Approach	Questions Answered
Consolidation	• Establish multiple call centers in a single baseline model • Establish rules for rerouting transactions to one or more centers • Test impact of rule changes on performance metrics • Adjust resources to support consolidation scenarios • Test different scenarios of call center consolidation • Find optimal solution based on goals and objectives of consolidation	• Should the call centers be networked or combined into a single center? • What is the optimal number of call centers in a networked environment? • Will agent performance levels improve if call centers are consolidated? • How much can staff and equipment costs be reduced? • How much are caller wait times reduced after consolidation?
Electronic Access	• Add inbound electronic transaction(s) to baseline model • Enter transaction volumes and call patterns for electronic transactions • Make adjustments to voice transaction volumes • Enter response times, priorities, routing logic and agents for electronic transactions • Test impact of electronic access on performance metrics at different volume levels • Select the most likely scenario (combination of voice and electronic transactions)	• What are the staffing requirements for electronic transactions? • What is the overall impact on staffing requirements for the call center? • What will the staffing requirements be if the electronic transactions grow at "x" percent in the next year? • How do electronic transactions change the cost structure of the call center? • How does this affect outbound transactions? • How do electronic transactions impact customer satisfaction?
Call Routing	• Establish baseline model of current operation • Establish new rules and conditions for routing incoming transactions • Establish agent groups based on skills and/or business knowledge • Simulate different combinations of call-routing rules and agent groups until model meets objectives	• How much can caller throughput be improved? • What is the impact on performance if call backs and call transfers are reduced? • What is the most appropriate use of call-center resources (for both efficiency and customer satisfaction)?

(continues)

Table 5.2. *(Continued)*

Trend	Simulation Approach	Questions Answered
Outsourcing	• Establish baseline model of current operation • Establish the transactions to be outsourced • Establish rules and conditions for outsourcing • Simulate different combinations of outsourcing rules until model meets objectives	• What is cost to outsource vs. keeping the function in-house? • Will outsourcing a particular function improve the overall efficiency of the operation? • How does outsourcing change the cost structure? • What impact will the outsourcing have on customer service?
Integration	• Establish baseline model of current operation • Modify the cost structure based on new hardware and software components • Modify call length, transfers, and call backs to reflect the improved efficiency and effectiveness of the operation • Modify agent groups to reflect skills and business knowledge • Modify incoming calls to reflect changes to electronic transactions and routing rules • Simulate different combinations of components until optimal solution is found	• What impact will "one and done" call processing have in terms of efficiency and effectiveness? • How will integration affect the performance of other functional areas in the company? • How will additional automation of existing processes impact cost and customer satisfaction? • What is the impact of adding new services to the call center?
Multifunctional	• Establish baseline model of multifunctional operations • Revise cost structure to capitalize on economics of scale • Establish new agent groups based on skills and business knowledge • Establish a single call center and route all multifunctional transactions into it • Create and test new routing rules • Find optimal performance of combined groups	• How does the new cost structure change the net revenue generated? • What is the impact of overflowing transactions to new agent groups? • How are performance measures impacted by combing multi-functional groups?

dressed by the model. While actual projects of this magnitude are much more complex, the approach is still the same. Simulation is one tool that all managers should use when undertaking major call-center redesign.

1. **Call-Center Consolidation**—Moving from a distributed to a centralized environment which can take the form of physical consolidation or networked centers.
2. **Electronic Access**—Moving from "call only" centers to providing customer access by any means the customer chooses, particularly electronic access.
3. **Call Routing**—Employing sophisticated call routing techniques to improve agent performance and customer satisfaction.
4. **Outsourcing**—Outsourcing one or more call center functions to a third-party vendor because of cost savings, a lack of internal resources, a lack of technical expertise, the need to implement a new program quickly, a lack of needed technology, etc.
5. **Integration**—Integrating hardware and software products that support new functions and improve customer access and service.
6. **Multifunctional**—Combining multiple business functions, such as sales and service, in a single call center.

REFERENCES

Purdue University Center for Customer-Driven Quality. 1999. *The 1999 Call Center Benchmark Report for All Industries.* West Lafayette, IN: Purdue University.

CHAPTER

6

Using Simulation as a Financial Analysis Tool

INTRODUCTION

Traditionally, call-center simulation tools have focused on improving call-center efficiency (i.e., finding new ways to reduce costs). However, not much attention (in simulation models) has been given to the revenue side of the equation. Spreadsheets have typically been used to analyze costs, revenues, and return on investment. Today, as call centers become more complex, the traditional analysis tools are becoming less effective. While managers use spreadsheets often for financial analysis and reporting, we suspect that very few would think to use simulation tools for the same purpose.

We aren't recommending that managers replace their spreadsheet applications with simulation tools. In fact, simulation tools and spreadsheets work very well together. What we want to demonstrate is how effective simulation tools can be in analyzing the call center's costs and revenues under a variety of conditions. Operating the call center at the lowest possible cost, without considering customer satisfaction, is likely to reduce or eliminate any positive return on investment.

BACKGROUND

Two critical benchmarks of a call-center operation are business efficiency and customer effectiveness.

> Spend too little and perform poorly, and your call center becomes a business liability that consistently drives away customers and creates market damage. Conversely, spend too much and over-perform, and your center again becomes a financial loss to the company. If you spend efficiently and perform effectively at a level just better than your competitors, your call center will most likely be a profit center for the company—i.e., getting, growing, and retaining customers. (Anton, pp. 1–4)

By putting these two dimensions into a performance matrix (Figure 6.1), we can compare the caller satisfaction index with the cost per full-time equivalent (FTE). Today, very few call centers are able to do both well. A recent study shows that only 10 percent of call centers are above average in both categories. How can these call centers take advantage of simulation to improve performance? Let's begin by defining the performance measures.

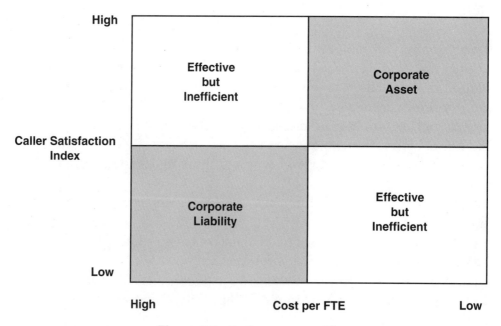

Figure 6.1. Performance matrix

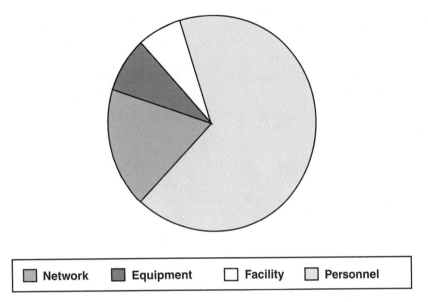

Figure 6.2. Call-center cost distribution

Efficiency

Most call centers focus only on improving the efficiency side of the equation. We measure efficiency by taking the total cost of the call-center operation and dividing it by the number of FTEs. Because personnel makes up approximately two-thirds of a call center's cost (see Figure 6.2), it is easy to see why call centers tend to focus on efficiency. With fewer people doing the same amount of work, operating costs are much lower. However, by focusing solely on efficiency and not on effectiveness with customer, it can actually end up costing the call center more. It is entirely possible for a call center to perform so poorly that customers become dissatisfied by virtue of the contact with the call center, thereby increasing the likelihood that they will disengage from the company's product or service at the next available opportunity. Remember the goal is to get the optimum level of customer service at the lowest possible cost. Focus on being slightly better than the competition without letting the costs outweigh the benefits.

Effectiveness

Being effective means delighting the customer into a continuous state of loyalty and satisfaction and a willingness to recommend the company's services and products to others. If a monetary value can be placed on customer effectiveness, then we are in a position to calculate the value the call center has to the business. By delighting the customers and being highly efficient, the call center can become an asset to the business.

Generally, to start making significant call center improvements requires an initial assessment followed by research, analysis, strategy, and eventually restructuring of the business. Simulation can be used very productively during the assessment and analysis phase, particularly when used as a "what-if" tool.

PERFORMANCE MEASURES

What information and performance measures can we expect a typical call center simulation tool to provide?

First of all, it is time based, allowing us to analyze the performance of the system (call center operation) over various time segments. Let's look at the example in Figure 6.3. We see an incoming call entering the model and going through a series of events (based on established rules and conditions in the call center model at that particular time). The model keeps individual call statistics on each event as it occurs. For instance, it calculates how long each call waits in the queue, how many times it transfers, which group the call was routed to, how long the call lasts, how many times it was put on hold, etc. Virtually anything that can happen to a live call can be captured in the model.

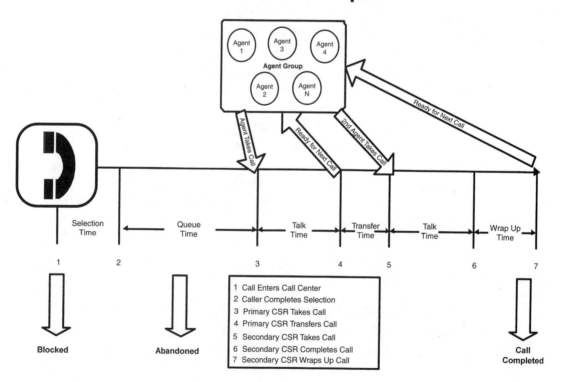

Figure 6.3. Call flow events

By summarizing all of the call activity over a specific period of time (½ hour, hour, day, etc.) the model can produce performance reports for that period. For instance, it could report the call center service levels in half-hour increments for a day, a week, or more.

In summary, the typical call-center simulation tool measures the internal performance of the call center based on the rules and parameters established in the model. This information, by itself, is not sufficient to measure the performance of the call center. We still need to know if the service level generated from the internal metrics is meeting the customer's expectations. This can be done by adding external (customer satisfaction) measures to the model. Let's walk through the steps.

EVALUATING THE CALL CENTER USING SIMULATION

Incorporating customer satisfaction measures into the simulation model is a four-step process. The first step develops the relationship between the customer's feedback and the internal metrics of the call center. It provides a gauge of where we are today. This is followed by calculating the costs and benefits generated by the customer contact to determine the call center's return on investment (ROI). Once performance measures, revenues, and costs are added to the model, sensitivity analysis can determine which metrics are the most sensitive to change. Finally, potential improvement alternatives are tested using a "what-if" approach. The four steps are:

1. Aligning caller satisfaction with internal performance measures
2. Calculating call-center ROI
3. Performing customer sensitivity analysis
4. Optimizing call-center designs

Align Caller Satisfaction with Internal Performance Measures

Unless the call center regularly surveys its customers, there is no way to know if the call center's current service level is meeting customer expectations. To do this, we link the external (customer) survey data to the internal metrics. Once this relationship is established we can test various scenarios designed to optimize customer satisfaction at the least possible cost. If the service level is set too high, the result is higher costs without substantial improvement in customer satisfaction. Conversely, if the service level is set too low, the center may be losing customers (and revenue). Let's look at the measures in more detail.

Internal Measures

Computers from your PBX (private business exchange), automatic call distributor (ACD), and voice response unit (VRU) generate call-center internal measures. These numbers are collected from the day-to-day operation and represent the recording of actual events in the call center. Examples would be how long the customer waited in queue or what was the average talk time. Other internal departments like accounting and marketing may also provide the necessary information to calculate such things as the total cost of the call center operation and customer lifetime value.

External Measures

External measures represent "soft" measures based on the perception your customers have at the time they take the survey. For instance, you may ask your customers to rate the length of time they waited before being connected to a CSR on a scale from 1 to 5 (with 5 being the most satisfied and 1 being the least satisfied). The customer's response is based on the satisfaction level with the service at this time. This is influenced by many factors (competition, past experience, demographics, etc.) and is likely to change over time.

If the call center is not conducting caller satisfaction surveys, there are other ways to get started. Call-center benchmark statistics are an excellent way to compare an operation to others in the industry and derive the initial external metrics. This should be followed by the call center's own customer satisfaction surveys. Both methods will be illustrated below.

Customer Survey

If the caller is surveyed immediately after the call is completed and the results of that survey are compared to the internal measures at the time of the call, a statistical relationship can be formed between the two. This relationship can then be used by the simulation model to establish a service level based on the customer's needs and expectations. Table 6.1 contains a list of customer survey questions and how they can form a relationship to the internal metrics of the call center at the time of the call.

In *Call Center Management "By the Numbers"* (Anton), the steps to form the statistical relationship are discussed in detail. Let's briefly review that process and show how it can be incorporated into a call center model.

$$y = B_0 + B_1 X + e$$

where:

y = dependent variable (the caller's satisfaction rating of queue time)
X = independent variable (the actual queue time)
e = random error (usually assumed to be 0)
B_0 = y-intercept of the line through the y-axis
B_1 = slope of the line

Table 6.1. Internal and external relationship metrics

Internal Metrics at Time of Call	Customer Perception (External Metrics)
# of Rings	Number of rings you heard before menu of choices was heard (scale of 1 to 10)
Queue Time	The length of time you spend on hold waiting for 1st CSR to answer (scale of 1 to 10)
Hold Time	Length time during call you were put on hold (scale of 1 to 10)
# of Transfers	The need for one agent to transfer you to a different agent to complete the call (scale of 1 to 10)
# of Transfers	CSR's knowledge about company's products and services (scale of 1 to 10)
Talk Time	Spent enough time with in handling your call (scale of 1 to 10)
First-Call Resolution	Did CSR answer your question or solve your problem during this call (yes or no)
Combination of Metrics	How satisfied were you with the overall service received (scale of 1 to 10)
Combination of Metrics	As a result of this call are you likely to continue your service with our company (scale of 1 to 10)

To illustrate, let's look at an example: A call center has recently implemented a random computer-assisted telephone (CAT) survey. It is conducted immediately after the customer completed the call. Each customer in the survey is asked the following question, "How satisfied were you with the wait time before a live agent answered?" The results of the survey are matched to the actual time (internal metric) the customer waited in the queue (see Figure 6.4). Once the survey is completed, the information is entered into a linear regression formula to calculate the relationship between the independent (time in the queue) and dependent (caller satisfaction) variables.

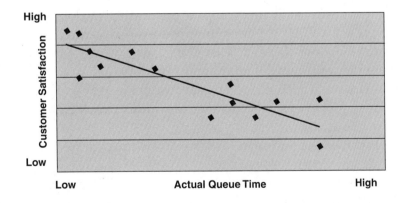

Figure 6.4. Relationship of customer satisfaction to queue time

The result is expressed in the following formula.

$$y = -0.554x + 91.53$$

It shows that for every second of improvement in the wait time the satisfaction rating improves by 0.55 points. If the average wait time is substituted into the formula, a customer satisfaction rating of 60 (please note that the customer survey responses have been recoded into a 100-point scale) is generated.

If we assume that a customer satisfaction reading of 85 and above is a delighted customer, an average queue time of 11.8 seconds or less is required to consistently meet that standard. To move from a satisfaction rating of 60 to an 85 requires a reduction in wait time of 45.1 seconds.

How is this information used in the simulation model? In the above example, the formula ($y = -0.554x + 91.53$) is added directly to the model. As each call leaves the queue, the model calculates the total wait time and substitutes this value for x in the above formula. The result is a wait time satisfaction rating for each call.

Over the simulation time period, the same calculation is applied to thousands of calls. The result is a call-center satisfaction rating for this external (wait time) measurement.

The next step is to find the relationship of all the individual measures to *overall* customer satisfaction. This is done by applying a regression formula, in a single equation, to all the attributes being considered. This formula would also be entered into the model. After the call is completed, the simulation model takes the calculated results of each individual metric and applies it to the multiple regression formula. The result is an overall satisfaction rating for each call generated by the simulation model.

Analyzing the Results

Table 6.2 is an example of the output a simulation model can generate from this type of analysis. Two new columns have been added to the report. They are labeled "Customer Satisfaction Index—Wait Time," and "Customer Satisfaction Index—Total." The information is summarized over particular time periods (as defined by the report). The same output could be generated for each and every survey question. With this new information, the model can *simulate* the customer's response to the internal metrics at the time of the call.

What is the model telling us? Are we delighting the customer or are we getting rejected because of poor service?

Our research shows that a score (index) below 50 represents a negative customer experience. In this case, we assume that the customer will disengage at the

Table 6.2. Customer satisfaction index

Time Slot	# of Calls	Avg Wait Time	Service Level	Handle Time	Customer Satisfaction Index	
					Wait Time	Total
1	30	10	90	3.1	92	78
2	45	5	87	4.5	89	76
3	124	35	78	4.1	76	67
4	145	65	74	5.3	55	35
5	121	76	76	4.1	49	65
6	89	45	84	5.6	34	62
7	79	76	65	6.6	41	55
8	110	110	55	7.8	25	45
9	55	15	89	4.3	74	79
10	44	8	91	3.2	88	85
11	65	21	88	4.2	79	90
12	23	4	98	4.6	93	45

earliest opportunity from the company's products and services. The company has lost the future revenue stream from this customer plus other potential customers from negative word-of-mouth.

If the score was between 50 and 85, the experience was neutral. The customer still uses the product or service but could leave at any time for a competitor. A neutral rating does not impact the future revenue stream.

If the experience was above 85, a positive outcome is generated. The customer has been delighted into continuing the relationship with the company. Value is added in terms of a future earnings stream both from the customer plus other potential customers from positive word-of-mouth.

The above output shows several time slots where the average wait time is high, causing customer satisfaction scores for that time period to be very low (meaning customer satisfaction is unsatisfactory). With this information in hand, a "what-if" analysis can be performed. But before we do that, let's look at another method (using benchmark data) to calculate customer satisfaction scores when customer survey data is not available.

Benchmark Data

If the call center has not developed a customer satisfaction survey program, benchmark data can serve as a substitute. For instance, Purdue University's Center for Customer Driven Quality takes an annual survey of call centers, collecting hundreds of valuable call-center metrics. This data can be used to derive quality of service metrics. The service metrics can then be applied in a formula to determine the call-center performance index. Let's look at the following example: A benchmark

Table 6.3. Using benchmark data to evaluate call center performance

Metric	Measure	Poor	Avg	High
Avg. Speed of Answer—ASA	seconds	30	20	15
Avg. Handle Time—AHT	minutes	7	3	2
Avg. Abandoned Calls—AAC	percent	10	5	3
Calls Handled on 1st Call—CHF	percent	30	75	90
Cost per Call—CPC	dollars	15	10	4

Call Center Perf Index = ASA + AHT + AAC + (100 − CHF) + CPC

Evaluation Scale	Scale
Asset	<50
Average	51–100
Liability	>100

report, for a particular industry, generated the following scores (see Table 6.3) for poor, average, and high levels of service. Other research indicated that these particular metrics were critical (have a high correlation factor) to customer satisfaction. Using this data, a formula was developed to calculate the call center's performance (see Table 6.3). The formula was input to the simulation model where it generated a rating for each simulated call. A rating of less than 50 was positive while those greater than 100 were negative. By feeding thousands of calls through the model, a call-center performance rating can be established by time period and in total.

Calculating Call Center ROI

Once customer satisfaction metrics (through customer surveys or benchmarking) have been added to the simulation model, we are ready to determine the call center's gross profit (loss). This is done by calculating the customer's value for each call received.

The last piece of information needed to calculate the call center's net profit and ROI (total net revenue/total operating cost) is the call center's total operating cost.

To summarize, the call center ROI is determined from the following information:

1. The customer's lifetime value
2. The value gained or lost as a result of the customer's contact with the center
3. The cost to operate the call center

Table 6.4. Customer lifetime value calculation (CLV)

CLV Formula	$R(1-(1/(1 + i)^n)/_i)$
Average Customer Lifetime (*n*)	4
Revenue (*R*)	$380
Interest Rate (*I*)	0.05
Customer Lifetime Value =	$1,347

Calculating the Lifetime Value of Customers

Let's again refer to Jon Anton's book, *Call Center Management "By the Numbers,"* to explain how to calculate the lifetime value (revenue stream) of the customer. When a customer is gained or saved through contact with the call center, we are, in fact, saving the present value of the customer's future revenue stream. Let's briefly illustrate the concept.

The present value of a stream of revenues from a customer can be expressed as follows:

$$CLV = R(1-(1/(1 + i)^n)/_i)$$

where:

CLV = present value of loyal customer's revenue stream

R = annual revenue received from a loyal customer

i = interest rate for the period

n = number of years of customer loyalty

If the formula is applied to the example in Table 6.4, we get a customer lifetime value of $1,347. This is based on an average customer retention of 4 years with a net revenue stream of $380 per year at a discount rate of 5 percent.

This is a very simple but important concept. Retaining loyal customers is extremely valuable to any corporation. We will demonstrate below how important it is for a call center to maintain and enhance this relationship.

The Value of Each Call

In step 1 we discussed how to calculate the customer's total satisfaction rating for each call coming into the call center. Based on the satisfaction rating, a value can be assigned to each customer's contact with the center as negative, neutral, or positive.

Table 6.5. Calculating call-center revenue

Customer X

Lifetime Value	$1,347
Overall Rating	91
Revenue Generated	
Gained/Lost Customer	$1,347
Positive/Negative Word-of-Mouth	$6,735
Positive/Negative Influence Factor	$ 67
Total Gain (Loss)	$1,414

Customer Y

Lifetime Value	$1,347
Overall Rating	64
Revenue Generated	
Gained/Lost Customer	$0
Positive/Negative Word-of-Mouth	$0
Positive/Negative Influence Factor	$0
Total Gain (Loss)	$0

Customer Z

Lifetime Value	$1,347
Overall Rating	45
Revenue Generated	
Gained/Lost Customer	($1,347)
Positive/Negative Word-of-Mouth	($26,940)
Positive/Negative Influence Factor	($539)
Total Gain (Loss)	($1,886)

Remember that a "delighted" customer is one who scores 85 or greater in the phone survey. This customer is assigned a positive rating. Conversely, if a customer scores below 50, we assume the business is lost. This customer is assigned a negative rating. If the customer score is between 50 and 85, a neutral rating is assigned.

Based on these ratings, Table 6.5 presents three examples of how the simulation model calculates call-center revenue. The model performs this calculation on each call.

The Total Cost to Operate the Call Center

The last component of our equation is the cost to operate and maintain the call center. The major call-center cost components are displayed in Figure 6.2. Let's assume that the total cost of operating the call center is $10,000,000 per year. The call center is operating 250 days per year with a full-time equivalent (FTE) staff of 100 customer service representatives. That brings the total annual hours to 200,000. Based on this data, the "fully loaded" hourly cost of a representative is $50. This hourly agent cost is entered into the model.

Call-Center Cost

Total Cost—Annual	$10,000,000
Days in Operation	250
FTEs	100
Total FTE Hours	200,000
Hourly FTE Rate	$50

Return on Investment

The last step in this process is calculating the call center's ROI. This is simply the revenue generated minus the total cost (sum of the hourly FTE cost) divided by the total cost. The model automatically maintains a running total of cost and revenue as each call is completed. Table 6.6 is a sample simulation output report incorporating both the cost and revenue components.

Sensitivity Analysis

Once the cost and revenue equations have been added to the model, we can begin performing a sensitivity analysis. Sensitivity analysis is a way to test the sensitivity of the model's output to small changes in an input parameter. In other words, it is a way

Table 6.6. Sample output that includes cost and revenue

Time Slot	# of Calls	Avg Wait Time	Service Level	Handle Time	Customer Wait Time	Satisfaction Index Total	Revenue	Cost
1	30	10	90	31.0	92	78	500	200
2	45	5	87	4.5	89	76	680	200
3	124	35	78	4.1	76	67	−400	350
4	145	65	74	5.3	55	35	−2,000	425
5	121	76	76	4.1	49	65	1,000	425
6	89	45	84	5.6	34	62	−2,300	400
7	79	76	65	66.0	41	55	500	400
8	110	110	55	78.0	25	45	−4,500	400
9	55	15	89	4.3	74	79	550	230
10	44	8	91	3.2	88	85	400	200
11	65	21	88	4.2	79	90	700	200
12	23	4	98	4.6	93	45	350	150

to determine which parameters will have the greatest impact on the desired measures of performance. Before this is done, be sure to confirm the model's accuracy.

Confirm Model's Accuracy

Any time a new model is developed or major changes are made, it's important to verify or re-verify its accuracy before the analysis phase begins. Depending on the complexity of the model, this could take a few minutes or much longer. The intent here is not to go through a detailed list of verification steps but to point out items that may be critical to an accurate representation of the call center model.

If the model is accurate, its output should be in sync with the "actual" metrics of the call center. Let's start with the major inputs to the model and compare them to the output reports generated from the PBX, ACD, IVR, and any other internal call-center report that reflects what is actually occurring in the call center.

Inputs

Incoming Calls	Staffing	Routing

For each call type, verify the following:
  Call volume
  Talk time
  Call arrival pattern
  Trunk group
  Average wait time before abandonment
  After-call work time
  Percent of calls having call backs
	For each agent group, verify the following:
	  Work schedule (start, stop, breaks, lunch)
	  Skill level by
	    Call type
	    Work priority
	    Agent proficiency
	  Call transfer percent
	  Conference percent
		For each routing script, verify the following:
		  Conditions for call routing
		  Built-in time delays
		  Queue allocation (single or simultaneous queues)
		  Call prioritization within a queue
		  Conditions for call flow between queues

Next, let's compare key outputs generated by the model to the statistics maintained by the call center. This would include the following measurements:

Output Statistics

- Percent abandon
- Average speed of answer
- Average talk time
- Average handle time
- Percent of calls blocked
- Average queue time
- Agent occupancy rate
- Agent utilization
- Percent handled on first call
- Service level

If these metrics do not agree with internal call-center reports, begin adjusting the model until it gives an accurate representation of the call center as it operates today.

Now let's look at the output from a different perspective. Does the model confirm the customer satisfaction ratings from the survey, both by individual attributes as well as in total? If not, go back and make sure the calculations are correct and all input data is accurate. Once the "external measurements" are confirmed, begin analyzing the actual results.

What Is the Output Telling Us?

Do we have the right combination of people, process, and technology to provide the greatest customer satisfaction at the least possible cost, i.e., the largest return on investment? What is the "service level" of this particular combination? What input variables or combination of variables have the biggest impact on the call center's costs and revenues? Ask these types of question as the sensitivity analysis begins.

In step 1 we discussed how to take internal and external metrics and define their relationship using simple regression models. We also determined the relationship of multiple individual attribute ratings to the overall satisfaction rating using a multiple regression model. One of the outputs from that model is a list of "independent" variables that have a significant impact on the dependent variable (overall satisfaction). If we know which variables are significant, we have our starting point for the sensitivity analysis.

Let's look at Table 6.7 on the following page. It represents data that was taken from a customer survey. The ratings have been re-coded on a scale from 1 to 100. If the data is entered into a multiple regression model, we get the following relationship:

$$y_1 = -13.35 + .554X_1 + .518X_2 + .099X_3$$

where:

y_1 = overall satisfaction rating

X_1 = wait time

X_2 = talk time

X_3 = number of rings

Intercept = –13.35

It shows that wait time is the most significant independent variable, followed by the talk time. With this information, let's go back to the simulation model's output report. Table 6.6 shows time slots 4, 5, 7, and 8 have average wait times much larger than the other time slots on the report.

Table 6.7. Survey results

Overall	Wait Time	Talk Time	# of Rings
44	44	56	89
33	44	56	44
44	33	67	55
78	67	89	100
56	67	67	33
67	56	78	56
67	78	67	22
56	67	67	56
89	78	67	78
44	33	56	33
22	33	22	78
78	67	78	78
67	78	78	67
89	89	78	56
44	56	44	78
33	44	33	33
78	89	89	78
56	44	67	78
88	78	78	67
33	22	44	56

If improving wait times has the biggest impact on overall customer satisfaction, then the question is, "What changes can be made to the simulation model to reduce these wait times?" While there are many potential answers, each case depends on the unique needs of the call center.

Let's assume, in the example, that more staff is the best solution. By adding more staff and rerunning the model, the impact to both the individual time slots and the overall model can be quickly determined. Look at the output and ask yourself, "Is the additional cost of adding new staff offset by improvements in customer revenue?"

To summarize, the process began by analyzing which variables have the biggest impact (most sensitive to change) on net revenue. In our example, it was wait time and talk time. To improve performance, we considered adding more agents to the time slots in question. Finding the best answer is an iterative process, which may involve several changes until the optimum design is found.

Optimize Call-Center Designs

All of our work up to this point has been to create an accurate representation of the call center as it operates today. We have shown the reader how to calculate the costs and revenues, add them to the simulation model, and analyze the results.

The end result is a very powerful analysis tool that can be used effectively to dramatically improve the operation. We are now able to input our ideas to the model and instantly see the dollar impact on the bottom line.

Effective call-center design is a very important step in maximizing resource efficiency while meeting customer needs. If the call center is not designed in the most efficient way, then it is going to be very difficult, if not impossible, for it to become an asset to the business.

The starting point depends on where the operation is today and the kind of changes being contemplated. Begin by asking lots of questions and developing a strategy for the future operation of the call center (in context with the total organization's goals and objectives).

1. Is the call-center design based on a particular customer relationship strategy or has it just grown over time without much thought to customer needs and expectations?
2. Are all customers who contact the call center treated equally? If they are, should they be?
3. Have customers been segmented by their value to the organization?
4. Are different strategies in place for different customers, and are the call center processes designed to support it?
5. What is the impact of these strategies on the efficiency and effectiveness of the call center?

The list of questions goes on and on. The right balance of efficiency and effectiveness will ultimately depend on the customer's needs. Once we know what it takes to delight the customer, the simulation model can help determine the most efficient operation to achieve that level of service.

In beginning to think about making changes to the call center—whether it's a major step like consolidation or just routine changes—simulation is a dynamic tool that can quickly evaluate the bottom line impact in terms of revenue and cost.

REFERENCES

Anton J. 1997. *Call Center Management "By the Numbers."* West Lafayette, IN: Purdue University.

Purdue University Center for Customer-Driven Quality. 1999. *The 1999 Call Center Benchmark Report for All Industries.* West Lafayette, IN: Purdue University.

CHAPTER
7

Dollars and Cents— Making the Case for Simulation at the Call Center

INTRODUCTION

We have demonstrated how simulation tools can be used to improve call-center performance. However, each call center has its own priorities, plans, and needs. Before considering the purchase of simulation tools, management should agree on what is to be accomplished, what the return on the investment is, and how it supports the business plan.

These and other questions will be addressed below so that software buyers can feel confident that they are making the right business decision. Remember, as with any technology investment, management of the investment is the key to realizing all the benefits.

Getting the most out of any software purchase requires careful analysis and planning *before* the software is purchased. Too many times expensive software is purchased and within a few months it's on the shelf collecting dust. Why does this happen? In many cases it is simply because the software is *"too difficult to learn and use.*" This happens because not enough time was spent analyzing the purchase and user needs.

Before the days of the personal computer with the graphical interface, software was primarily judged on its functionality (i.e., could it perform all of the business functions and provide the necessary output?). While that is still important, a major criterion today is the software's "usability." If it isn't intuitive and easy to use, people no longer have the time or are willing to take the time to learn it. The best software vendors today spend a great deal of time and money on the user interface. It is something that decision makers cannot afford to overlook and must be given the proper weight during the evaluation process.

Before purchasing, the software buyer must be willing to do three things to ensure that the right decision will be made: 1) do careful research, 2) talk to others, and 3) try it out. Our experience shows that talking to others who use the software on a regular basis is one of the best ways of getting past all the claims made by the vendor selling the software. Most references are very forthright and honest in their appraisal of the software.

Before an investment decision is made, management must decide if simulation tools make sense based on the call center's current state and plans for the future. Let's start the process with the following question.

DO SIMULATION TOOLS ADDRESS THE PROBLEMS AND NEEDS FACING TODAY'S CALL CENTER?

If a call center is having the type of issues or problems outlined below, then simulation should be given strong consideration.

1. The call center is growing very rapidly.
2. Major upgrades to the technology and call-center processes are in the planning stages.
3. Call-center agents may begin doing outbound sales as well as continuing to support inbound customer calls.
4. Consolidating multiple call centers is a key consideration.
5. New call centers are being added to support the increased volume of calls.
6. A recent benchmark study shows costs and service levels are well below others in the industry.
7. Customers constantly complain about the service they receive (the wait time is too long; they have to call back several times; calls are transferred too often, etc.).
8. A new call-center manager will need to determine where the call center is in terms of efficiency and effectiveness.

9. Several new products are being added to the call center.
10. Workforce management software is available, but it is difficult to use and does not seem to be very accurate at predicting staff assignments.
11. Outsourcing components of the call-center operation is in the planning stages for the next year.
12. Turnover is very high.
13. Plans are under way to route calls to the most qualified customer service representative available.

These types of changes reflect many of the critical issues facing call centers today. Many are experiencing high growth with new technology, high turnover, new products and services, and new choices for servicing the customer, and have a customer base that expects more and better service.

If a call center operation is facing similar issues, then the following questions should be asked.

1. Are the customers surveyed on a regular basis?
2. How does the operation compare to the competition and to call centers in general?
3. Is the call center well integrated with other parts of the business in terms of work flow and customer touch points?
4. Has a benchmark study or a comprehensive assessment of the current operation been done in the last year?
5. Are there a clear set of goals and measurements, and is the call center meeting or exceeding them?
6. Is the call center in the top 10 percent in terms of efficiency and effectiveness?

If the answers to these questions are not clear or no steps are being taken to improve the operation, then it may be time to consider simulation tools. On the other hand, if the call center is already "best of class" or few if any changes are expected over the next year and the current results are satisfactory, simulation tools are probably neither needed nor justifiable.

Simulation tools work best in an environment that is dealing with the types of issues outlined above—an environment where change is the norm and the decision making process typically involves multiple alternatives that are complex, high risk, and have a high dollar impact. During a project's analysis phase, these tools are very good for documenting and communicating information about the current processes. They are also very good at predicting the performance characteristics of proposed processes and providing a guide for future process performance.

HOW WILL SIMULATION TOOLS BE USED?

If the initial evaluation shows that simulation tools do, in fact, have the potential to identify improvement opportunities, we have passed the first hurdle in the evaluation process. Now we can begin addressing more detailed questions about how the tool will be used.

Who Will Use the Simulation Tool?

When the evaluation process begins we need to know who will be responsible for developing and analyzing the models. Will it be call-center personnel or a support group such as IS (information services)? If there are multiple call centers with their own staffs, will the tool be shared or belong exclusively to one group?

An important consideration is choosing the right individual or individuals who will be responsible for all simulation development. While the tools are becoming easier to use, they still require someone who has strong analytical skills and understands the workflow and call routing procedures.

What Are They Going to Use It For?

What call center projects are planned that can use simulation tools to support the analysis phase? Where and how can simulation tools be used to help solve the biggest problems currently facing the call center? All the short- and long-term projects to which simulation can be applied must be identified. These projects will be used to support the justification process.

Where Is It Going to Be Used?

Will the tool reside in a particular call center or in a central location to be used by several call centers? Will the software reside on one PC or be used by multiple PCs operating on a network? The answers to these questions also enter into the justification process.

When Will It Be Used and Needed?

This goes back to the set of problems and issues (the actual projects) that will be addressed using simulation. Will the software be used on both a routine and an ad hoc basis? How many groups will be using it? By thinking through when simulation will be needed and determining the size of the projects involved, the time and resources required to complete the projects can be estimated.

Why Will This Type of Tool Be Used?

In an earlier section of the book, we compared simulation tools to other analysis tools, listing the pros and cons of each type. For each potential project the user should determine if a particular simulation tool is the best fit for solving the problem.

VENDORS AND THE SIMULATION TOOL MARKETPLACE

After the above questions have been answered and there is a feeling of confidence that the need for simulation tools has been established, a detailed evaluation of vendors and tools can begin.

Let's take a closer look at the simulation tool marketplace by asking the following questions.

What questions need to be asked about the marketplace for simulation tools?

What questions should the vendor be asked?

How is the software evaluated?

What features and functions are critical?

What to buy: a tool or consulting?

Simulation Tool Marketplace

While there are exceptions, the simulation tool marketplace consists of many small vendors. These vendors typically deliver capabilities in a niche area, have relatively low revenues associated with their tools, and do not offer a front-to-back life cycle solution. (By life cycle solution, we mean tools that perform analysis, design, development, and the implementation of a given solution.) Most simulation tool vendors provide the analysis portion but do not provide development or implementation of the results of the analysis.

If we think of process analysis tools, we generally include diagramming tools, workflow analyzers, and simulation tools. Some vendors integrate these three components while others have a single-purpose tool that may integrate with other vendor packages. For instance, a simulation vendor may use Visio or FlowCharter as the diagramming tool for its product. The key here is a well-integrated product that has seamless connectivity to other products that need to work with it.

What does this marketplace mean to the potential buyer? Is there more risk associated with acquiring this type of software from a small niche vendor? What does it mean from a long-term perspective? Below is a list of considerations that need to be researched about the vendor before making the decision. Determine if the vendor's direction fits with your plans and future needs.

Key Questions about Vendors

General Information and Financials

The goal is to determine the financial health and viability of the company, both now and in the future. Learn as much as possible about the product being considered for purchase, such as how many customers they have, growth rate, research and development expenditures, etc. Where do they expect to be three to five years from now? Who are their key competitors?

System Support

Find out what kind of software support they offer. Is it 24/7 (24 hours, 7 days a week), and is there an additional charge for the support? What items are included in the support costs provided?

Vendor Commitment

Ask when the latest version was released. Find out when the last four or five versions were released. A major release can be expected every 12 to 18 months. Look at product release notes to determine what was in the last version or two. If no major enhancements are indicated and there is a long period between releases, the vendor is probably not committed to the product.

Documentation and Training

What training materials are available? What kind of training is offered and can it be tailored to individual company needs? Can it be done on site and what is the fee? How is training priced? What documentation comes with the product? Are user manuals included in the cost of the product? If not, how are they priced?

Consulting Services

Does the company offer consulting services? What is the range of the services they offer and how is it priced?

Pricing/Contract/Warranty

How is the product priced? Does the size of the call center (number of seats) affect the price? Are there volume discounts? Is there an annual maintenance fee and how is it priced?

System Advantages

What advantage does the simulation software from this particular vendor offer in comparison with competitor products?

References

The vendor should be able to provide a list of as many as 10 companies that are using the software. Ask for specific names of persons you can talk to at these sites. Prepare a list of detailed questions and call as many as you can. Ask what they think about the product and the company. Find out what their experience has been with technical support. The information from this exercise will be invaluable.

Evaluating the Simulation Software

Two key considerations in choosing software today are learning time and productivity. The user of the software must be productive very quickly. While software has continued to make great strides in ease of use (on-line help, graphical front ends, better manuals, etc.), over the last 10 years, subtle differences in software still make some packages much more difficult to learn than others.

If a software package is difficult to learn initially, then it is probably going to be difficult to relearn to use later. Ask the question (of yourself and others): "If I don't use the software for a week or two, how easy is it to quickly get back up to speed?" Remember that many call centers do not need to use simulation software every day.

How do we know if one vendor's software is easier to learn and use compared to others? To answer this question it is necessary to spend time doing the proper amount of research. Remember, choosing software means judging how well it actually works. Just because the software has the features and functions needed does not mean it works well or is easy to implement. It is important to choose software than will grow with the call center and its changing demands.

Below is a list of tasks that will provide valuable information about the software:

Read Reviews

Unfortunately buying software is not like buying a car. It's not possible just to look at it, walk around it, look at a manual, and make a decision. Sometimes it is very difficult to see the things that are missing.

Try the Software

Get a trial version of the software and spend some time trying it. Watching an expert demonstrating software always makes it look easy until someone tries it out back at the office. If possible, try to develop a small model so that the software can be used productively even in the trial.

Determine Simulation Method Used

Take care to make sure that the tool chosen incorporates discrete event simulation, and not analytical simulation. Some call center simulation vendors use analytical simulation techniques by incorporating some forms of randomness and variability by carrying out modified Erlang calculations in very small time increments to create an *illusion* of discrete event simulation. Some of these tools also refer to the type of simulation they use as "call-by-call simulation." Many such tools also have a strict spreadsheet orientation; hence they appear to be easy to use. These characteristics also make them easy to recognize. Due to the nature of the analytical calculations, such tools also provide extremely high-run speeds. However, the results from such tools can be dangerously misleading and not necessarily better than Erlang calculations. In order to test whether a certain product or vendor is actually doing discrete event simulation (looking at the true dynamic nature of the system, at each time advance), ask the following questions about a simulation tool and example run.

- Is it possible to find out the status of the system at 8:05:23 A.M. or P.M. (or at any other specific point in time during the day)? (If information is obtained for time slots beyond the point chosen for the status report, it is likely that the tool being used is an analytical calculator—which is not a discrete event simulation tool.)
- Can the path taken by a call or transaction be followed from start to finish?
- Can the status of a system (such as global settings, variables, call attributes) be changed at any point in the run cycle?
- Can the logic be stepped through one time advance at a time?
- Can information in all time slots (from the beginning of the day to the end of the day) be filled and complete results obtained—even if the run was just started and it ran for a millisecond?
- Can minimum, maximum, and average values of specific calls be figured by the tool?
- Can information be gotten on every run?
- Can the status of a particular call be checked at any point in time?
- How can I make sure that my dynamic rules are being obeyed (rules such as routing based on minimum expected queue times, average speed of answer, etc.)?
- Will the tool show how the time advances and how the system state changes at every time advance (which are the guts of discrete event simulation)?

Research the Market

Find out what tools are the most popular and why.

Talk to the Experts

Talk to consultants and users whose jobs depend on how well the software works and how easy it is to use.

Find out about the support offered. Software should be enabled with support mechanisms that are both formal (training and manuals) and informal (Internet discussion groups, user groups, e-mail lists, etc.).

Ask about the level of interoperability. The simulation software must get along with other software in and out of the call center. Look at the inputs and outputs to other systems. Getting data into the model and output into another system can become an important task. Are there industry-standard formats that are followed?

Determine if the user can extend the features of the software. What templates and analytical extensions are included in the software? This is very important because work changes. Software needs to respond to these changes. For instance, can the software be extended to include an Internet component? If the software cannot respond to global changes, it becomes ineffective. Software must be capable of rapid response to new advances, which are coming at a much faster pace then just a few years ago. It is impossible to wait for next year's release to respond to last year's problems.

Find out if the software is reliable. Are there frequent problems with the software not working properly or "blowing up" during a run?

Make a List of Important Features and Functions

These are discussed in more detail in the following section.

Develop an Evaluation Table

Develop a list of features, functions, and vendor information that are critical to the call-center operation. Apply a weighting factor to each item based on its relative importance. As each software tool/vendor is reviewed, rate it according to the criteria established. This rating will provide a relative comparison of the products reviewed. A sample evaluation can be found in Table 7.1.

What Features and Functions Should Be Considered?

General-Purpose versus Call-Center-Specific Tools

The software we are evaluating is a time-based (dynamic), discrete event modeling tool for call center analysis. A discrete event model means the variables (calls and

Table 7.1. Simulation software evaluation example

Requirement	Weight	Product 1	Product 2
Simulation Methodology (Call-by-Call vs. Discrete Event Simulation)	5	3	1
Ease of Use	5	2	4
System Support	4	3	5
Training	4	5	3
Product Price	3	2	4
Consulting Services	4	2	5
Documentation	5	5	3
Vendor Long-Term Viability	5	4	3
Software Extensibility	5	3	5
Model Capacity	4	3	5
User Help	5	5	4
Planning Horizon	3	4	3
Outbound Calling	5	2	4
Input Transaction Types	5	2	3
Skill-based Routing	5	5	5
Cost and Revenue Features	4	2	3
Variables and Equations	3	3	3
Simultaneous Queuing	5	5	5
Input Distribution Analysis	2	2	4
Entity Attributes	2	2	3
Animation	2	3	4
Reporting—Ad hoc	4	2	4
Reporting—Built In	5	5	3
Software Interface—Input	3	0	4
Software Interface—Output	4	4	4
Network Requirements	1	2	5
Hardware Requirements	2	4	4
Hierarchical Modeling	4	0	5
Total Score		**328**	**409**

other transactions entering the call center) change instantaneously and at separate points in time, i.e., the number of calls in the system changes only when a new call arrives or when a call is completed or terminated. These tools come in two categories: general-purpose and those designed specifically for call-center modeling. While the underlying simulation engine is the same, those designed for call centers already have the call-center components and reports built into the model. This is very important because the model development time is substantially reduced.

User Interface

As we discussed earlier, the ability to quickly learn and be productive with the software tool is a very important consideration. This is closely tied to the user interface and how it was designed. Today's simulation tools should have a graphical (GUI) interface using industry-developed standards. Since this type of simulation normally depicts tasks and the flow between them, it should allow the use of standard flowcharting tools such as Visio or FlowCharter. Building the model should be basically a "point-and-click" operation with all the necessary components already available. If formulas are required, they should be point-and-click type entries. The interface should be intuitive for the user who already understands the call-center components and how they should integrate into a call-center model.

Extensibility

How can the features of the software be extended by the user? The software should be capable of modification and extension to meet future needs. One example is the question: "Can the software be extended to support Internet transactions or will it be necessary to wait for the next release?" Another example would be: "Can the transaction be extended to include other actions that occur outside of the call-center environment?"

Capacity

What is the capacity of the modeling software in terms of number of seats and transactions entering the call center? The question of capacity is also likely to be dependent on the processing power of the PC. Simulation software can require a large amount of memory. Processing can also slow considerably as the size and transaction volume of the call center increase. The vendor should have statistics available on the capacity of the modeling software.

User Help

It is very important that user manuals and on-line help be complete, easy to understand, and contain examples of how to build certain types of models and take advantage of everything built into the tool. Because models can become complex, examples are very important to learning and productivity. Without good documentation, model development time can increase exponentially.

Hierarchical Models

In most cases, the first-cut model describes the process in general terms. If activities require further detail, submodels of that process can be developed. Without the ability to create hierarchical models, the modeling process can be come very complex and difficult to manage.

Distribution Fit

Incoming calls, call-processing times, and other activities generally have patterns that can be represented by probability distribution functions. Distribution functions eliminate the need to use actual input. The software should have the capability to take actual input and determine the distribution function from it.

Trouble Shooting

Because of the underlying complexity of simulation models, errors can be generated when the model attempts to compile and run. If the software is designed properly, it can catch many user errors in advance of compilation. As an example, the software can have input panels to enter formulas that make it very difficult, if not impossible, to make a syntax error. The more of these built-ins, the better. If an error does occur, the error message and the ability of the software to display where and how to fix it can save many hours of frustration.

Animation

Visualization provides the context that is often needed to understand the collection of events that make up the process. With a graphical interface, a visual rendering of the work-flow and call-flow process can be displayed. With animation, the process "comes alive" as the model runs. This is useful for validation and verification of the call-center model.

Product's Ability to Flowchart Call Flow

Being able to model an ACD's routing design isn't so useful if it's not possible to quickly and easily verify that the model is working properly, and, furthermore, that exactly what's happening in the ACD has been captured. Good products give this critical "eyes-on" capability with flowchart animation. Flowcharting is an ideal intuitive tool that allows users to graphically map the complex routing of calls between call centers and agent groups. Flowcharts "come to life" as icons representing

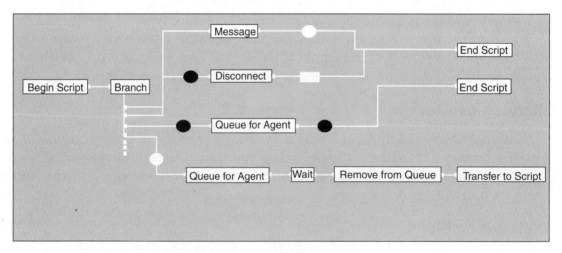

Figure 7.1. Call flow example

each call "flow" through the logic blocks and the connection lines in the flowchart. These fundamental capabilities to quickly map ACD logic enable easy verification and validation of its accuracy. The ability to toggle this feature off and on, depending on the requirements of the end-user, should be available. Figure 7.1 is an example of what a typical call-handling process would look like. This enables easy validation of the call-routing logic and determination of where the bottlenecks may be occurring.

Cost and Revenue

If the simulation tool provides the ability to enter costs and revenues, the model can generate key financial data for building the business case. Most simulation tools are weak in this area. Ideally, the software will provide an easy-to-use interface to enter all cost and revenue component data. In particular, look for the following types of costing analysis:

- **Agent costs:** Ability to describe fixed agent costs in terms of hourly wages. The aggregation of costs incurred by agents is then calculated automatically as the simulation progresses based upon the schedule followed.
- **Trunk costs:** Ability to describe trunk costs in terms of costs incurred per minute. As the simulation progresses, costs incurred for using trunks are automatically calculated based on trunk utilization.
- **Fixed/Variable costs**: Ability to customize any costs associated with lost calls, costs per call, and revenue opportunities per call, and to tie in any mathematical formulas associated with the costing analysis. Both fixed and variable costs can be modified through relationships that tie in several factors such as changes in service levels, abandonment rates, revenue collected, cost of service, and so on.

The combination of the three methods enables you to perform a complete financial assessment of your call center, from budgeting to capital investments and resource allocations.

Output Data Analysis

The software should allow the user a great deal of flexibility on reporting results. Typically, simulation tools provide a standard output report of key activities and measurements. Most provide the ability to generate graphs and charts. Call-center simulation tools should allow the user to report all of the standard call-center metrics in addition to generating custom reports to fit individual needs. Tools should also provide the option to create the output in commercial spreadsheets such as Excel.

Scenarios

This feature allows the user to set up and run a batch of simulations and analyze them at a later time. The software should provide the tools to evaluate and compare each test case.

Attributes and Variables

Attributes are values associated with individual entities. Attributes can be used to make decisions, perform calculations, etc., during the simulation run. For instance, an incoming call could be assigned a customer attribute and then, based on that value, be routed to a particular agent group. Variables are also place holders for values. They are global in nature and can be used for keeping count on the number of calls waiting in a queue, calculating the average wait time, etc. The ability to assign and use attributes and variables in a model allows the user the flexibility and capability to develop a model that behaves like the real-world process. The ability to assign attributes and variables in a model should be part of any good simulation tool.

Interfaces to Other Software Systems

Is the tool able to interface directly with the PBX/ACD to get incoming transaction data? Can the output from the simulation model be fed into other software products such as spreadsheets or planning tools?

Hardware, Software, and Networking Requirements

What are the hardware and operating software requirements for running the simulation software? Will the software run in a network environment where multiple developers can work on a single model?

Should Outside Consulting Be Considered Instead of a Simulation Tool?

Before putting the business case together, another alternative to consider is using outside consulting for simulation needs. This approach can work very well, for instance, if a project fits the criteria for simulation, but there is no interest in the purchase of a simulation tool.

Before choosing outside consulting, first determine the scope and objectives of the engagement. Is outside consulting needed for a look at other aspects of the call-center operation or should the focus only be on developing simulation models? Most simulation software vendors offer consulting specifically to develop models and train internal personnel. Also, many business-consulting organizations use simulation tools during the analysis phase of the project. Generally speaking, consultants should be considered in the following situations:

- **Time constraints**—Simulation models are needed quickly, and there is no one available internally who has either the time or the skills to complete the request within the planned time frame.
- **No in-house expertise**—The user does not have the in-house expertise to do the analysis and develop the simulation models.
- **Objective opinion**—The user wants a third party to either confirm the analysis and modeling already completed or develop other alternatives based on his or her particular knowledge and expertise.
- **Training**—The user recently purchased the software and wants initial or advanced training for the in-house staff. In many cases, a specific project is developed jointly by the in-house staff and the consultant, giving the in-house staff the opportunity to learn simulation outside the classroom.

DEVELOPING THE BUSINESS CASE

We are now ready to take all the information that has been developed and make a compelling and credible case for the justification and purchase of a simulation tool.

The business case must support the planning and decision making process, and, in particular, it must support decisions about whether to buy the tool (or use outside consulting), which vendor to choose, and when and how to implement it. It also presents the financial benefits of buying and implementing the simulating tool to management. It will show the expected cash-flow consequences of the decision over time and will include the rationale for quantifying benefits and costs. If relevant, critical success factors and significant risks should also be presented. In addition, the business case should also include a discounted cash flow, payback period, and internal rate of return.

Below is an outline of the key components that should be part of the formal business case presentation.

A. Introduction and Overview
B. Assumptions and Methods
C. Business Impacts
D. Sensitivity, Risks, and Contingencies
E. Conclusions and Recommendations

Introduction and Overview

This section contains relevant information about the author or authors, when and how the business case was developed, and a clear statement that describes what the case is about. The statement describes the planned action (purchase of simulation software) and objectives that will be accomplished as a result of this action. Included in this section is an executive summary (subject, scope, methods, results, and financial metrics), which states as concisely as possible the purpose and conclusions of the study. The introduction provides background, overview, and a description of the situation leading to the purchase decision.

Assumptions and Methods

This section describes the assumptions and methods, the scope and boundaries, and the cost/benefit model used to explain where the data and results came from. It describes how the financial case was built and the metrics used (such as return on investment, payback period, cost per customer, cost per employee, etc.).

The assumptions section should include the factors used to project future financial results and clarification of costs and revenues (examples: the average cost of service representatives or a customer's lifetime value).

Discussion of the scope of the analysis should include the time period planned for analysis, the sites involved, the specific simulation projects and how they will be implemented, which organization and functions are involved, and the hardware and software required to complete the project.

The cost/benefit model includes all the items developed during the analysis. The data sources and methods used to develop the model should also be included.

Business Impact

This section makes the business case. It answers the essential questions about what the financial consequences will be if the proposed action is taken. Generally, this would be a cash-flow statement showing all the line items that represent true cash inflows and outflows. This statement can be used to develop an internal rate of return and the return on investment figures. Also included is an analysis of the financial model and the cash flow results.

Sensitivity, Risks, and Dependencies

The purpose of the sensitivity analysis is to measure the sensitivity of the results if the assumptions change. This exercise examines changes in the model's primary output based on changes to individual input factors. The types of factors that could impact this type of analysis are salary levels, incoming call volume, new products, new technology, lifetime value of customers, turnover, productivity levels, customer satisfaction measures, etc. This section should demonstrate which assumptions could have the biggest impact on the financial results if the assumption is wrong. The analysis should include upper and lower ranges on the inputs to give the reader a complete picture of the range of possibilities.

Risk analysis attempts to analyze the risk of the financial results in terms of probability. How likely is the projected financial statement compared to the likelihood of other financial results? Dependencies outline what must be done and by whom and in what time frame in order to fulfill the expected financial results. It puts everybody's commitment on record, thus helping to ensure the goals will be met.

CONCLUSIONS AND RECOMMENDATIONS

This section succinctly summarizes the complete case using supporting evidence from the preceding sections to demonstrate the author's reasoning. All the important decision criteria necessary for the case should be stated in this section. The conclusion section should focus on the expected contribution of each objective developed earlier. The recommendation restates the conclusion based on the preceding analyses, bringing closure to the case and a shift in the responsibility for action to those who will make the final decision.

Completing the business case will provide confidence that a clear need for simulation tools has been established, the right tool and the right vendor have been selected, and that a plan is fully in place for achieving the goals set forth.

Case Studies

1

SBC Communications

Bob Bushey, Senior Member of Technical Staff, SBC Communications

SBC Communications, Inc., is a global leader in the telecommunications industry, with more than 36.9 million access lines and 6.5 million wireless customers across the United States, as well as investments in telecommunications businesses in 10 other countries. Under the Southwestern Bell, Pacific Bell, SNET, Nevada Bell, and Cellular One brands, SBC, through its subsidiaries, offers a wide range of innovative services. Among them are local and long-distance telephone service, wireless communications, data communications, paging, Internet access, messaging, telecommunications equipment, and directory advertising and publishing.

What Prompted the Use of Simulation Technology at Your Corporation?

Southwestern Bell Corporation (SBC) considers numerous changes in our call centers on a continual basis. These changes are focused on improving the operational environment in one aspect or another. Each of these proposed changes needs to be evaluated as to its respective costs and benefits. Obviously, each proposed change needs to have the benefits exceed costs of that change. In many cases, we were unable to perform this "cost/benefit" analysis with the degree of confidence that was necessary to justify the amount of expenditures and the potential impact on our customers. Most of the tools that were at our disposal (and, to our knowledge, of the industry) were not capable of evaluating call-center operational complexities.

One of the critical call-center performance measures is the customer satisfaction metric. How well we serve the customers and their perceptions of that service continues to be an important aspect in today's business climate. An evaluation of the customer's total experience based upon the average experience does not fully value the individual circumstances of each and every customer. For example, if the average time it takes to answer a customer's call is 10 seconds, that may not be a cause of great concern. However, if that 10-second average represents some customers who are experiencing 45-second delays, their negative experience may remain hidden. Those "long delay" customers could be expected to have low satisfaction and may be hidden from exposure, thus hidden from a solution. So, a tool which has the ability to consider the actual occurrences of customer events (rather than averages) is an important aspect of evaluation for us. Simulation technology offered us a solution to resolve such issues.

What Were the Specific Advantages Provided by Simulation Technology?

One of the most important simulation capabilities for SBC was the ability to consider individual customers and their respective experience with our call center. SBC's focus is on the customer, and many other technologies lack the capability to, in a detailed sense, consider the total customer experience. One of the risks of call-center management is evaluating performance from a set of averages. The customer is never an average, his or her experience is a unique experience, and the better we are able to evaluate that individual perspective, the better we can serve our customer.

The ability to evaluate the complex call-center operational environment and the complex proposed changes to that call-center environment was a distinct advantage of simulation technology. Most proposed changes to the call-center environment have unexpected ramifications to other segments of the process. It is far better to acknowledge these impacts before the change is implemented than to have operations "discover" the impact when the actual customer is affected.

What Other Complementary Technologies Were Used for Call-Center Analysis?

Spreadsheets are probably the most common tool that is used in many organizational units. This type of tool is excellent for stable, sequential, non–time-dependent processes. The reality is that the call-center environment is not stable, many processes are parallel, and the activities are certainly time dependent. Other tools provide valuable insight, but our experience indicates that those tools do not consider sufficient factors to be valid for analysis in and by themselves. Simulation technology was used in conjunction with these tools to provide a powerful "what-if" analysis tool that enabled us to make business decisions.

What Types of Problems and Issues Did Simulation Help You Resolve?

SBC is constantly considering improvements to our operational environment in order to improve customer service. As in many organizations, SBC wished to consolidate some of our individual, relatively small call centers into larger, "virtual," consolidated call centers. Individual service representatives would not have to physically relocate, but by linking diverse smaller call centers, a higher level of service can be provided. Of course, management wanted to be sure that these types of consolidation would improve customer service before we went through the time and expense of the change. SBC created simulation models that enabled us quickly to manipulate different scenarios and conditions for consolidation. Furthermore, they provided us with the confidence and the correct degree of accuracy required to facilitate consolidation.

From a more detailed perspective, simulation technology appears to be unique in assisting the evaluation of situations where there are multiple customer calls arriving at the same time (but the arrival rate changes by hour of day, day of week, day of month, and month of year) with multiple agents handling those customer calls (and agent schedules varying by shift, day, and vacations). This creates a complex interaction resulting in a dramatic variation of call duration.

How Would You Quantify the Return on Investment from Using Simulation Technology at SBC?

The most significant benefits have been to our customers. We have been able to implement many process improvements because we have been able to evaluate, with confidence, the various proposed changes via a verifiable cost/benefit analysis. We have been pleased with the simulation technology and believe we have demonstrated that this tool has been worth the time and effort of designing, building, and running the simulation models.

Do You Have Any Recommendations for an Organization That Is Just Starting Out with Simulation?

One recommendation that probably stands above others is to explicitly incorporate in the project plan-specific "simulation technology" assistance in order to get the project off to a quick start. If the project team is new to simulation technology, then hire outside assistance to streamline some of the "blind alleys" that may be encountered. If the project team has some experience with simulation technology, then purposefully plan on an aggressive kick-off where a preliminary model with some operational data can be available quickly. Simulation technology, as its corresponding animation, resonates so well with upper management that it is important to quickly have a running model in order for the project team to solidify their support.

2

Navy Federal Credit Union

Eddie Pruitte, Call Center Analyst, Navy Federal Credit Union

Navy Federal Credit Union is the world's largest credit union, with over $10 billion in assets and more than 1.7 million members. Headquartered in Vienna, Virginia, it has 85 member service centers and 203 "no surcharge" ATMs worldwide. The credit union serves most military and civilian personnel of the Navy and Marine Corps and their families. The union offers a myriad of financial consumer products and services to its members 24 hours a day, seven days a week. These products include, but are not limited to, auto loans, mortgages, credit cards, and electronic/home banking. Through its 6 Number Concept, its 12 largest call centers are a primary source of member contact. Last year Navy Federal call centers answered in excess of 35 million calls.

What Prompted the Use of Simulation Technology at Your Corporation?

By definition, call centers are unique and dynamic in their operation. Several components of telecommunications technology and human engineering must successfully merge together and function properly if the call center is to be successful. With technology changing daily and the unpredictability of human resources, a tool that provides the flexibility to consider these elaborate parts and their varied individual characteristics was needed. Simulation was such a tool.

What Were the Specific Advantages Provided by Simulation Technology?

Unlike any other tool I have been introduced to, simulation allows you to duplicate real-life scenarios before implementation—a sort of desktop laboratory. Most decision-makers do not enjoy being taken by surprise, especially when the results are not favorable. Simulation also gives you the ability to consider the intricacies of different telephone calls and how they are delivered, processed, and concluded.

Efforts to deliver world-class service have resulted in many call centers developing complex environments. As call centers broaden their horizons and abilities to satisfy customers, call types and their associated characteristics have become complicated as well. As the call types become more departmentalized, the dynamics of the call center change and, subsequently, all of these variables must be considered as one makes recommendations.

What Other Complementary Technologies Were Used for Call-Center Analysis?

We use Merlang and the typical spreadsheet applications (Lotus and MS Excel). While those tools may suffice for "quick and dirty" analyses, results (outputs) are given in averages. The concepts of randomness, spikes, uniform distribution, and other statistical phenomena are absent.

Making decisions on the assumption that every call will last the same amount of time, or each agent works at the same skill level, or the arrival pattern of calls does not vary tremendously limits the decision maker's ability to make an intelligent, accurate decision. For example, a loan application call may consume twice the handling time of a change of address call. Both call handling times are distinct and should be treated so. The skill levels of agents varies based upon their experience and ability. To place all agents in the same category is not only questionable, but also could lead to inaccuracy.

What Types of Problems and Issues Did Simulation Help You Resolve?

We use simulation predominantly for staffing and capacity analysis. Without question, the most significant cost for a call center is human resources. As such, having the amount of staff in the right place at the right time is vital. Too many agents means unnecessary salaries and benefits, which decreases the profit margin. Having too few agents could mean less revenue-generating calls. Both scenarios are to be avoided at all costs.

Simulation has also been utilized to assist senior management in determining when to add phone lines or determine the peak capacity of our interactive voice response (IVR) equipment. We have also used simulation to maximize our human resources, that is, deploying skills-based routing more effectively.

How Would You Quantify the Return on Investment from Using Simulation Technology at Navy Federal Credit Union?

Call-center managers have used simulation outputs to strategically place agent resources where they will have the greatest impact. We have also used simulation to determine the requisite facilities to meet demands for our automated services.

Do You Have any Recommendations for an Organization That Is Just Starting Out with Simulation?

Be prepared for the task of assembling data. The adage "garbage in, garbage out" is undoubtedly overused, but I honestly cannot think of a better term. The user must know what data is available and how it should be gathered, input, and interpreted before the simulation results will have trustworthiness.

The user should also consider building baseline models and comparing the outcomes to historical results. This affords the user an opportunity to determine how accurate his or her model is and builds credibility among his or her customers/audience.

It is inevitable that you will also have to make some assumptions. Make sure your audience understands them and they accept the assumptions. If not, the simulation outcome will be questioned.

Do not become overwhelmed with analyzing data (also known as "analysis paralysis"). Know when you have gathered enough data to construct the simulation model. Also, do not assume results before the simulation is concluded. This tendency has a way of impacting the data input and interpretation.

In conclusion, remember the simulation model is only as good as you construct it. If good data is gathered and placed in the model properly, you will have more confidence in the results. Discuss your methodology with someone as you construct the model. Be open to suggestions and input from people unfamiliar with the simulation concept. If they can understand your approach and techniques, chances are others will and you will become more believable.

3

Communications Data Service, Inc.

Eric Jacobs, Industrial Engineer,
Industrial Engineering Department, CDS Corporation

Communications Data Services, Inc., (CDS) is an international data management company. Our clients are an elite group of the world's largest magazine publishers and direct marketing firms. In business since 1972, CDS is backed by the stability and experience of our parent company, The Hearst Corporation. CDS provides order fulfillment services for more than 380 accounts, including national magazine titles and direct marketing firms. CDS maintains data representing the names, addresses, and demographic information of over 105 million active magazine subscribers—more than one-third of the U.S. population. CDS is the "connection" between our clients and their customers. CDS receives orders—either over the phone, through the mail, or via the Internet—on behalf of magazine publishers and direct marketers. Once those orders are entered, the customer information then resides on one of CDS's computer systems. The information is available for clients to access for their own analysis and business purposes. CDS uses the information to mail magazines, products, bills, and renewals and to handle customer service.

What Prompted the Use of Simulation Technology at Your Corporation?

We felt that simulation was the only tool to use that would enable us to see what was happening inside our automatic call distributors (ACD) in a visual format. It also would allow us to try out different skill-based routing changes without having "trial by error" occurring with live calls.

What Were the Specific Advantages Provided by Simulation Technology?

The most valuable advantage that we saw involved the animation in the routing structure. With different routing strategies, we were able to see the results visually as they would occur in real life. It gave us a better idea of what would happen with a changing environment, so we're better able to respond to the changes.

What Other Complementary Technologies Were Used for Call-Center Analysis?

We used MS Access heavily as an additional tool with our simulation software. The database was used to collect and store ACD statistics that we later used in supplying data to the different models.

What Types of Problems and Issues Did Simulation Help You Resolve?

Our simulation models helped us better route calls to agents that are more familiar and skilled to address the customer's needs. It also helped us see where we needed more or less staff at the right times during the day. Basically, simulation helped us do more with less.

How Would You Quantify the Return on Investment from Using Simulation Technology at CDS?

We found immediate benefits from the first day of implementations resulting from our simulation study. These included improvements in key call-center performance measures, including a reduction in call waiting times, shorter agent talk times, and fewer staff needed to achieve our objectives.

Do You Have Any Recommendations for an Organization That Is Just Starting Out with Simulation?

Use simulation as another tool to complement analysis techniques within your call-center environment, but exploit the abilities you have within all of your tools.

4

IITRI

Katherine Miller, Senior Computer Scientist, IITRI

IIT Research Institute (IITRI), founded in 1936 and affiliated with the Illinois Institute of Technology, is one of the country's largest not-for-profit contract research institutes. IITRI is working with government and industry clients to find solutions to information management challenges and enterprise modernization needs by bringing to bear expertise in information systems security certification, acquisition planning and management, electronic commerce, advanced data analysis, communications engineering, information system architecture development, performance measurement, simulation and modeling, and data management.

IITRI is currently supporting a large government agency in implementing intelligent call routing capability and centralized workforce management for their customer support division. IITRI engineering support has included modeling and simulation of the agency's call-center environments for use in economic analysis and justification of information system investments consistent with today's congressionally mandated use of performance measures and portfolio investment techniques.

What Prompted the Use of Simulation Technology at Your Corporation?

The government agency currently provides toll-free 800 numbers (referred to as product lines) for customers to call for product-related questions. When a customer calls any one of the toll-free numbers, the customer can get routed to one of 25 call centers located throughout the U.S. that are staffed with customer service representatives (CSRs) belonging to agent groups that are trained to handle specific types of calls. During peak season, the agency employs about 6,000 CSRs.

The government agency installed a GeoTel Intelligent Call Router® (ICR) at 9 of the 25 sites on a trial basis to test its call routing technology as an alternative to its current system of routing. The Call Router is a commercial off-the-shelf system that continually monitors and receives status messages from call-center locations. The Call Router uses data that it receives (e.g., queue waiting times, CSR idle time, etc.) from call centers in order to decide where an incoming taxpayer call should be routed. The routing decision is based upon routing logic that can be dynamically modified.

Up until this time, the agency had used a routing technology based on work-force percentage allocation. The percentage allocation method of routing directs a percentage of the total number of incoming calls to call centers based on the projected number of CSRs that are scheduled to work.

The prototype was initiated to test the Call Router technology in the agency's environments and to provide data to support a business case for a national roll-out of the Call Router should customer service decide to proceed in that direction. Baseline sites that routed calls by percent allocation were selected to provide performance data for comparison to the nine Call Router prototype sites. However, several factors have had an impact on the comparative analysis of the two competing routing methodologies:

1. The prototype operated with low call volumes (about 15% of the normal call volume) during a non-peak period. As a result, it would have been impossible to extrapolate what would happen during a full-scale implementation of the Call Router during a peak filing period.
2. At least during the period that the prototype was being tested, it was determined that management practices at the different call centers were having a significant impact on productivity and thus skewing the results of the baseline sites versus prototype sites' performance comparison.
3. During the period of prototype testing, major script changes were being evaluated. For example, changing the routing algorithm from "most available agents" to "longest available agent." Major script changes would, to some extent, render previous prototype results obsolete.

We became interested in the use of simulation because it provided a means to perform a comparative analysis of the two competing routing methodologies without the real-life factors that would influence or skew the results.

What Distinct Advantages Did You See in Simulation Technology That Piqued Your Interest?

We were interested in the following advantages that modeling and simulation offered:

1. Simulation eliminates other factors that influence performance. For example, when we run two simulations, one for percent allocation and one for Call

Router, we can create equal conditions for each simulation run: the same volume of calls, same call arrival pattern, same number of agents to handle calls, etc. In addition, factors such as site outages, hardware failures, or trunk failure, which influence performance in real life, can be controlled or eliminated with modeling.

2. Changes can be made easily to the model to reflect Call Router script changes.
3. It is cost effective to model the entire environment. Due to cost considerations as well as risk factors, the prototype was only fielded at nine sites and only applied to one product line at each site.
4. Extrapolation is supportable due to the ability to model light and heavy volume days.

What Were the Specific Advantages Provided by Simulation Technology?

Aside from prototyping that was used in the business case analysis, we also investigated using the GeoTel ICR® Lab System. The Lab System uses the actual GeoTel hardware and software of their call center Intelligent Call Router, which is augmented by other hardware and software to support receiving simulated calls on a call-by-call basis. As such, the ICR Lab System is a real-time emulator. There were two overriding factors that limited the use of the GeoTel Lab System for the business case analysis:

1. *Cost and acquisition time:* It was estimated that the Lab System would, under the best of circumstances, require a lead time of at least a month to acquire services, necessary hardware, and perform installation. The Lab System requires at least five Pentium processors and extensive computer memory to simulate the number of subject sites (25).
2. *Real-time call processing:* Because the Lab System processes call records in real time, the number of experiments that can be performed are time constrained. A standard-day experimental run requires 24 hours to process, whereas a standard-day experimental run using discrete event simulation can be executed in minutes. Hence, discrete event simulation allowed many and various experiments, whereas only a few specific experiments could have been conducted with the Lab System due to overall schedule constraints.

Another factor that was considered was that the use of the GeoTel Lab System to evaluate competing routing technologies could be perceived as bias.

What Types of Problems and Issues
Did Simulation Help You Resolve?

Simulation was used for business case benefits and return on investment (ROI) projection. We constructed a model of the existing call-center environment at the government agency using percentage allocation routing to send calls to call centers. We then modeled the same call system dynamics and intricacies under the Call Router system. Subsequently we conducted average-day simulations with light and heavy volumes and other "what-if" laboratory analyses and experiments to facilitate planning decisions that are required to be documented and substantiated in the business case.

How Would You Quantify the Return on
Investment from Using Simulation Technology at IITRI?

The first benefit was the ability to get results that were accepted by the agency's customer service department as realistic and well substantiated. Business cases, which are developed in order to justify an information technology (IT) investment, are often challenged because they claim benefits and cost savings that are not well supported.

Any Other Information That You May Want
Your Peers to Know About Your Experience?

The modeling and simulation enabled the agency to estimate a Call Router operational cost per call under four categories in increasing call volume. It was interesting to note that the projected savings from using the Call Router were reduced as call volumes increased. The simulation validated findings that the agency had observed on a limited basis. As a result, the agency is now considering alternating between both routing technologies based on projected call volume. That is, to switch to percentage allocation routing on days with extraordinarily high call volume.

Do You Have Any Recommendations for an
Organization That Is Just Starting Out with Simulation?

A popular nationally syndicated radio talk show host often attributes his popularity to the idea that people like to listen to him because he validates what they think and has the ability to articulate their beliefs, and he uses his show to further expand on those ideals. We can draw a parallel between the attraction of the popular talk show and simulation.

In our case, we were pleased because the simulation validated what we had postulated. But its most significant benefit was that it provided the details (i.e., busy agent cost, trunk costs, number of calls serviced, etc.) of how much of a benefit the investment would return. That enabled us to better articulate the benefits and provide specific IT investment measurement objectives. It also went beyond what we had initially intended to ascertain from simulation by helping us to understand under what conditions we could expect such a return and how factors such as incoming call volumes can have a significant effect on our operational cost per call.

With today's complex information systems, modeling and simulation are no longer "nice-to-do" options. It would be unconscionable to make any significant network, hardware, software, and staffing investments for such information systems without getting confirmation of results from modeling and simulation efforts. Business case, ROI, technical viability, choke point, and system limitation analyses of information systems permit the assessments of a large number of complex interactions that cannot be evaluated by simple closed-form methods.

5

Oracle Corporation

Sam Kraut, Business Planning Specialist, Oracle Support Services

Oracle Corporation is the world's second-largest software company and the leading supplier of software for enterprise information management. The company has two major businesses: one aimed at providing the lowest cost information technology infrastructure and the other to provide business and competitive advantages through high value applications. With annual revenues exceeding $7.5 billion, the company offers its database, tools, and applications products, along with related consulting, education, and support services, in more than 140 countries around the world.

Oracle Support Services is the industry's leading provider of software support for information management systems. The Support Services organization has more than 6,600 experienced and knowledgeable professionals currently supporting more than 150,000 customers, more than 150 Oracle products, and more mission-critical systems than any other software company.

Oracle's support infrastructure provides global, around-the-clock support. Global support centers in the United States, the United Kingdom, and Australia offer customer support 24 hours a day, seven days a week. Oracle Support Services also has more than 90 local support centers worldwide that provide support in local languages.

Why Did You Consider the Use of Simulation as One of the Alternatives in Your Analysis Toolbox?

Oracle Support maintains five major international call centers, linked to provide global around-the-clock service. Evaluating their efficiency and predicting their performance required a sophisticated tool that enabled us to model the intricacies of our support environment.

We feel that complex interactive situations are best modeled and refined through simulation before committing the resources in real life—an aircraft manufacturer does not roll a new design right off the assembly line, load it with passengers and crew, and send it up to see whether it flies! Just as aerospace engineers use flight simulators to test new designs, call-center managers and analysts can use

117

call-center simulators to test and refine organizational changes quickly and cheaply before trying them out on their actual agents and customers. In fact, simulation is the only method that can realistically model the dynamic relationships between a call center's organizational plan, staffing plans, and call routing scripts for detailed analysis.

What Types of Problems and Issues Did Simulation Help You Resolve?

Oracle Support used simulation primarily to evaluate a proposed reorganization in the call handling process. This study included replicating the actual call routing scripts in the phone system and incorporated mathematical algorithms that mimic the probabilistic nature of call arrival rates, talk times, abandon behavior, etc. In our opinion, this represented a big improvement over spreadsheets, which were inadequate for many types of analysis. Second, simulation was used to project the headcount required to maintain a prescribed service level and responsiveness criteria as dictated by management.

What Were the Specific Advantages Provided by Simulation Technology?

We were able to achieve the first goal of our simulation, that is, projecting the headcount required to achieve a certain level of service by tracking reports that included average hold time, maximum hold time, and the percentage of calls that held for less than a specified length of time. Our simulation models replicated peak load behavior and the random nature of real call arrivals to show how service levels can vary during the day. The models also helped us better understand our processes and point out areas that were especially sensitive to change.

How Would You Quantify the Return on Investment from Using Simulation Technology at Oracle?

Simulation technology enabled us to determine with confidence the pros and cons of a proposed reorganization in our call handling process and better understand our staffing deployment. In addition, there were several unexpected lessons learned from modeling the center. First, it pointed up the serious implications of abandons. When abandon rates are high, some customers are getting served at the expense of others, even though average response times might be relatively low. Second, the model showed the importance of keeping outgoing and direct-to-agent calls to a minimum. Third, the model showed a clear tradeoff between skill-based routing and quick response times. Overspecialization among agent groups can cause degradation in service.

Having built a basic model of our center "as is," the next step was to construct models of hypothetical call centers based on different organizational plans. Specifically, we were evaluating the impact of changing from a two-level to a one-level response arrangement. Under the two-level plan, dedicated response teams answered the calls and handed off technical issues to separate resolution teams. In the single-level model, all agents would answer calls and resolve issues. Line managers had concerns that agents in a one-level plan would get swamped, but the model showed that even if call volumes were much higher than projected, agents would have ample time to do both call response and "backlog" work.

Is There Any Other Information That You Want Your Peers to Know About Your Experience?

In our opinion, a simulation tool should play a central role in the business planning process for any call center. Once overall strategy has been determined for the enterprise, call-center management can focus on which operating tactics will best realize the corporate strategy. For example, if around-the-clock coverage is required, should this be implemented via 24/7 staffing at one center or a "follow-the-sun" strategy across multiple centers around the globe? Simulators can prove useful in evaluating such alternatives. The model builder gathers a variety of information from the planning process and information systems, including call volume forecasts, incoming call patterns, seasonality, call lengths, agent scheduling, and call routing scripts to construct both "as is" and hypothetical models. Output figures for gross staffing requirements are fed into a scheduling optimizer to fine tune the scheduling plan. Simulation modeling can be an iterative process, with revised models taking into account schedule refinements and service level goals. Final output on staffing requirements becomes input for the human resources department's hiring plans and the financial department's budgeting process. The result is a continuous, integrated planning process from strategic direction to budgeting and financial results.

6

Bank of America

Charles T. Maner, Vice President,
NationsBank Marketing Group/Channel Research and Analysis

BankAmerica Corporation is a global financial services enterprise with offices in 38 countries, a capital base of more than $20 billion, and assets of more than $260 billion. This global strength and range makes us a world leader in corporate financial services. As a result, thousands of corporations, institutional investors, financial institutions, governments, and central banks around the world rely on BankAmerica Corporation as a powerful financial ally.

Bank of America, with $618 billion in total assets, is the largest bank in the United States, with full-service consumer and commercial operations in 22 states and the District of Columbia. Bank of America provides financial products and services to 30 million households and two million businesses, as well as international corporate financial services for transactions in 190 countries. Bank of America stock is listed on the New York, Pacific, and London stock exchanges and certain shares are listed on the Tokyo Stock Exchange.

What Prompted the Use of Simulation Technology at Bank of America?

Simulation provides a means of evaluating and experimenting with a system, in this case a calling center network, through the use of a computer model rather than on the system itself. Prior to simulation there were few tools that could adequately encapsulate the variability and randomness inherent within a calling center. Also, due to different calls being answered by different agent groups within routing scripts, closed form equations, such as the traditional ErlangB and ErlangC, made simulation an obvious alternative. Our study was done to lend support to the activities associated with the Direct Banking Planning and Operations group at Bank of America.

What Were the Specific Advantages Provided by Simulation Technology?

Simulation allows the analyst a wide range of freedom to experiment with various scenarios. The analyst can try a myriad of different routing scripts and call-handling scenarios, thereby replicating the logic of the actual call script or call router used at the call center.

Simulation can also be a unique demonstration tool. Not only do most simulation packages offer animation, but they all produce statistics directly comparable to statistics used to measure the performance of the real call-center network. Management throughout the calling center planning process can therefore understand these measures that simulation produces—they are identical in nature to traditional measures used within the actual calling center environment.

From a scenario perspective, the analyst may want to evaluate the effects of different schedules on service levels by time interval. Additionally, simulation can indicate discrepancies in routing logic that may otherwise not be apparent through current, traditional measures. Likewise, the analyst can point these areas out to call-center managers as well as recommend changes to alleviate the discrepancy. At present, we know of no tool other than simulation that allows the analyst a quick answer to these questions.

What Other Complementary Technologies Were Used for Call-Center Analysis?

There were really no competing tools or technologies to consider. We were committed to simulation as a tool and were left only with deciding which simulation package to purchase. As for complementary technology, we used optimization tools for scheduling and forecasting tools for forecasting since simulation is not intended to do either. The only additional tools for this simulation study were TCS Management Group, Inc., for scheduling and agent information and MS Excel for synthesizing the information obtained from the various tools.

What Types of Problems and Issues Did Simulation Help You Resolve?

The primary problem that we, Bank of America, are currently working on is determining operating hours from a cluster of calling centers that handle similar call types. There are three calling centers in a cluster, one of which is a 24/7 site and two other non-24/7 sites. The goal is to set operating hours such that the closing of one, or both, of the non-24/7 sites does not "slam" the 24/7 site at the time of closing with call volume such that staff scheduling is grossly inefficient.

In order to address this planning problem, we have developed a simulation completely replicating the call routing logic, call patterns, average handle times, and average after-call work times for call types. The result is a simulation model with which we can evaluate the effect of different closing times on the 24/7 site. Standard, quantifiable measures will be customer handling times and customer service levels by time period, as well as agent utilization by time period.

As new centers are added to a cluster, or if we want to incorporate different clusters simultaneously, the process is relatively straightforward. Therefore, should we want to evaluate multiple centers' operating hours on customer service levels, simulation provides that capability.

Additionally, we plan to evaluate different operational scenarios. These scenarios include evaluating the effectiveness of different skill-based routing scripts, customer service levels of using cross-trained agents vis-à-vis cost to appropriately train those agents, priority queuing, and ongoing evaluation of agent schedules.

How Would You Quantify the Return on Investment from Using Simulation Technology at Bank of America?

The benefits from employing simulation technology are dollar savings. These dollar savings are achieved from a quick turnaround of analyses, tangible dollar savings resulting from simulation studies, and intangible savings through knowing that a good, analytically based solution was found.

Through employing simulation, we have a much clearer understanding of the ramifications of system changes, such as operating hour changes, among our centers. Furthermore, analyses can be done fairly quickly without having to design a suite of experiments and then perform those experiments within our actual centers.

Do You Have Any Recommendations for an Organization That Is Just Starting Out with Simulation?

We advise that the analyst have a clear understanding of the problem to be studied. Also, make certain that the analyst and the clients clearly understand and agree on what measures will be used for evaluating different scenarios. It is advisable that a measure be adopted such that each scenario can be quantified and ranked accordingly. Depending on the particular simulation study, these measures may be simply cost or some cost–customer service level measure, that is, a unit increase in customer service equates to some cost increase.

The clients must also have and understand the data needed to perform an adequate simulation study. More often than not, without proper data a simulation study cannot come to a confident conclusion.

There must also be a substantial amount of client interaction. Without an adequate knowledge of the environment or question being simulated and studied, the simulation study will result in little more than an academic exercise for the analyst. The client must be involved in identifying when an acceptable solution has been found and what tradeoffs in one solution versus another solution are acceptable.

Any Recommendations for Other Users?

About the only additional advice we have, given that simulation is adopted as an analytical tool, is to hire staff with an appropriate background and training in simulation. Developing a simulation is more than simply coding. It involves developing a user agreement of what is to be modeled via simulation. If a user agreement is too formal for the client-development team, an unambiguous description of what is to be studied and assumptions regarding the modeling approach need to be clearly articulated. Also, the individual must have an understanding of what data is needed and how to address inevitable data gaps. The individual should have an understanding of how to analyze simulation results, both from a standpoint of verification and validation, as well as drawing statistical inferences from simulation output.

Simulation models use probabilistic/mathematical components to simulate the actual performance of an operation. The appropriately trained analyst understands any limitations that may exist between the mathematical model and the actual operation and takes them into account during the study, thus maximizing the chance for success.

7

United Parcel Service

I Ding, UPS Business Communications

United Parcel Service (UPS) Business Communications Services (BCS) is a wholly owned subsidiary of United Parcel Service of America. UPS is the world's largest express carrier and largest package delivery company, serving more than 200 countries and territories. Headquartered in Atlanta, Georgia, UPS has been recognized for the past 16 years as *Fortune Magazine*'s most admired company and operates one of the largest and most sophisticated global telecommunications networks today.

BCS draws on the vast technical expertise, reputation for reliability, financial stability, and global presence of UPS to develop the solutions that will deliver value and satisfaction to its customers. BCS currently handles inbound, outbound, and blended customer calls, and Internet communications; manages bilingual and international call services; administers forecasting and scheduling; collects and mines customer data; and provides customized reports and customer care consulting. Hundreds of thousands of calls are handled each day in their 10 customer care centers located throughout the country.

What Prompted the Use of Simulation Technology at UPS?

BCS constantly evaluates the latest technological tools to best serve our customers. Call simulation software was chosen for its ability to predict the future needs of our customers. Because of our past data-mining activities and the vast amounts of data captured, the use of call simulation software seemed like a natural extension of the BCS planning activities.

What Were the Specific Advantages
Provided by Simulation Technology?

The main advantage was the ability to quickly evaluate our customers' business situations by performing "what-if" analysis to identify their strengths and weaknesses. As a provider of strategic consulting services in call center and customer care management, simulations offered BCS the ability to combine cost, scheduling, and simulation together to determine the most cost-effective way to provide quality service to potential and current customers.

What Other Complementary Technologies Were Used for Call-Center Analysis?

BCS is currently evaluating Call Center Staffing Tools, GeoTel's Intelligent Call-*Router,* and Computer Telephony Integration systems.

What Types of Problems and Issues Did Simulation Help You Resolve?

With call volume consisting of sales, service, billing, information, emergency, bilingual, and international calls, BCS operates one of the most dynamic call-center environments there is. When operating multiple call centers and servicing the varied needs of multiple customers, simulation becomes a vital tool in determining the most cost-effective way to maintain quality and utilization. Simulation assisted BCS in analyzing the number of subgroups that a single call center or multiple call centers can operate effectively. Simulation also allowed us to experiment with segmentation models to improve the quality of our customers' calls.

How Would You Quantify the Return on Investment from Using Simulation Technology at UPS?

Simulation technology allows BCS to expand our technical expertise and offer our customers reliable service and financial stability. Simulation technology complements BCS's strategy of assisting businesses with the development, management, and maintenance of their customer service systems by providing them with customized solutions to satisfy their customers' needs. The ability to determine each customer's baseline and the total impact that each customer has on the overall call center was a benefit provided by simulation technology.

Do You Have Any Recommendations for an Organization That Is Just Starting Out with Simulation?

Developing an advanced simulation requires capturing large amounts of data and the complete understanding of your call-center processes. Before considering the use of simulation as a tool, ensure that you have the resources and a complete understanding of the goals you are attempting to accomplish. In addition to having the data available to support your simulation scenarios, you should establish a system to validate the simulation results. At BCS we found that benchmarking is vital when implementing simulation results and models and ensures that you do not incur any negative financial results due to the use of an improperly validated model.

Any Recommendations for Other Users?

When considering any simulation tools, ensure you not only research the product but also the company behind it. Technical support, product flexibility and stability, and company experience may be a few of the elements that make your simulation initiatives a success.

Index

Page numbers in *italics* denote figures; those followed by "t" denote tables.

About the Authors

Jon Anton, PhD, MS, BS

Jon Anton is a researcher at the Purdue University Center for Customer-Driven Quality. He specializes in enhancing customer service strategy through inbound call centers and teleweb centers using the latest in telecommunications (voice), and computer (digital) technology, as well as the Internet, for external customer access along with the Intranet and middleware for organizing and delivering company information now stored in limited access databases and legacy systems.

Dr. Anton has assisted over 400 companies in improving their customer service strategy/delivery by designing and implementing inbound and outbound call centers, as well as by assisting in the decision-making process of using teleservice providers for maximizing service levels while minimizing costs per call. In August of 1996, Call Center Magazine honored Dr. Anton by selecting him as an Original Pioneer of the emerging call-center industry.

Dr. Anton has guided corporate executives in strategically repositioning their call centers as robust customer access centers using a combination of re-engineering, consolidation, outsourcing, and Web enablement. A single point of contact for the customer results from this repositioning and it allows business to be conducted anywhere, anytime, and in any form. By better understanding the customer's lifetime value, Dr. Anton has developed techniques for calculating the ROI for customer service initiatives.

Dr. Anton has published 48 papers on customer service and call-center methods in industry journals. In 1997, one of his papers on self service was given the "Best Article of the Year" award by the magazine *Customer Relationship Management*. Dr. Anton has published five professional books: 1) *Call Center Management "By the Numbers,"* published by Purdue University Press, 1997; 2) *The Voice of the Customer,* published by Alexander Research & Communications, 1997; 3) *Customer Relationship Management* by Prentice-Hall, 1996; 4) *Inbound Customer Call Center Design* by Purdue University Press, 1994; and 5) *Computer-Assisted Learning* by Hafner Publishing, 1985.

Dr. Anton's formal education is in technology; he holds a Doctorate of Science and a Master of Science (Harvard University), a Masters of Science (University of Connecticut), and a Bachelor of Science (University of Notre Dame). He has also completed an intensive, three-summer, Executive Education program in Business at the Graduate School of Business of Stanford University.

For questions, this author can be reached at 765-494-8357; drjonanton@aol.com.

Vivek Bapat, MBA, MS, BS

Vivek Bapat is the product marketing manager for the Arena Product Line at Systems Modeling Corporation. Systems Modeling Corporation is the developer of Arena®-based products and a world-leading supplier of simulation technology and professional services.

Prior to this position, Mr. Bapat was product manager for Arena Call Center, an award-winning simulation solution for call center planning and analysis. Under Mr. Bapat's guidance, Arena Call Center received 1998 Product of the Year awards from *Call Center Solutions Magazine* and Call Center News Service. It also received an Editor's Choice Award from *Call Center Magazine* and *Call Center Focus Magazine* in the UK.

Mr. Bapat has over ten years of broad experience in the simulation field, including the development of special-purpose simulation solutions, consulting, customer support, sales, and marketing, with special focus in the services industry. In these roles he has assisted several leading companies across the globe in improving their customer service strategies to achieve world-class service through the use of simulation.

In addition to conducting numerous conference presentations and seminars on the value of simulation in improving business processes, Mr. Bapat has also published several articles and papers in industry journals and national publications.

Mr. Bapat received his MBA from Robert Morris College in 1997, his Masters Degree in Industrial Engineering from Clemson University in 1991, and his Bachelor of Science in Mechanical Engineering from the College of Engineering in Pune, India in 1988.

For questions, this author can be reached at 412-741-3727; vbapat@sm.com.

Bill Hall, MBA, BS

Bill Hall is a management consultant with Call Center Services, where he assists clients in improving their call center operations and in re-engineering major business processes. Mr. Hall's specialty is in using computer simulation tools to analyze call-center performance and to develop strategies supporting management decisions that drive call-center operations. Mr. Hall has worked in positions in senior information technology management for over 20 years; his industry experience ranges from manufacturing and distribution to the insurance and health care industries.

Mr. Hall has a strong background in systems analysis, application development and design, project management, and strategic planning. He received his Masters Degree in Systems Management (Baldwin Wallace College) and his Bachelor of Science in Industrial Management (Purdue University).

For questions, this author can be reached at 440-257-0642; bhall@mailbag.net.